Advanced Strategies in Thin Film Engineering by Magnetron Sputtering

Advanced Strategies in Thin Film Engineering by Magnetron Sputtering

Special Issue Editors
Alberto Palmero
Nicolas Martin

MDPI • Basel • Beijing • Wuhan • Barcelona • Belgrade • Manchester • Tokyo • Cluj • Tianjin

Special Issue Editors
Alberto Palmero
Spanish Council of Research (CSIC)
Spain

Nicolas Martin
FEMTO-ST Institute, University of Bourgogne Franche-Comté
France

Editorial Office
MDPI
St. Alban-Anlage 66
4052 Basel, Switzerland

This is a reprint of articles from the Special Issue published online in the open access journal *Coatings* (ISSN 2079-6412) (available at: https://www.mdpi.com/journal/coatings/special_issues/film_Magnetron_sputter).

For citation purposes, cite each article independently as indicated on the article page online and as indicated below:

LastName, A.A.; LastName, B.B.; LastName, C.C. Article Title. *Journal Name* **Year**, *Article Number*, Page Range.

ISBN 978-3-03936-429-9 (Hbk)
ISBN 978-3-03936-430-5 (PDF)

Cover image courtesy of Raya El Beainou, Ph.D. Thesis 2019, Univ. Bourgogne Franche-Comté, France.

© 2020 by the authors. Articles in this book are Open Access and distributed under the Creative Commons Attribution (CC BY) license, which allows users to download, copy and build upon published articles, as long as the author and publisher are properly credited, which ensures maximum dissemination and a wider impact of our publications.
The book as a whole is distributed by MDPI under the terms and conditions of the Creative Commons license CC BY-NC-ND.

Contents

About the Special Issue Editors vii

Alberto Palmero and Nicolas Martin
Advanced Strategies in Thin Films Engineering by Magnetron Sputtering
Reprinted from: *Coatings* 2020, *10*, 419, doi:10.3390/coatings10040419 1

Amine Achour, Mohammad Islam, Iftikhar Ahmad, Khalid Saeed and Shahram Solaymani
Electrochemical Stability Enhancement in Reactive Magnetron Sputtered VN Films upon Annealing Treatment
Reprinted from: *Coatings* 2019, *9*, 72, doi:10.3390/coatings9020072 7

Jan-Ole Achenbach, Stanislav Mráz, Daniel Primetzhofer and Jochen M. Schneider
Correlative Experimental and Theoretical Investigation of the Angle-Resolved Composition Evolution of Thin Films Sputtered from a Compound Mo_2BC Target
Reprinted from: *Coatings* 2019, *9*, 206, doi:10.3390/coatings9030206 15

Florian G. Cougnon and Diederik Depla
The Seebeck Coefficient of Sputter Deposited Metallic Thin Films: The Role of Process Conditions
Reprinted from: *Coatings* 2019, *9*, 299, doi:10.3390/coatings9050299 29

Manuela Proença, Marco S. Rodrigues, Joel Borges and Filipe Vaz
Gas Sensing with Nanoplasmonic Thin Films Composed of Nanoparticles (Au, Ag) Dispersed in a CuO Matrix
Reprinted from: *Coatings* 2019, *9*, 337, doi:10.3390/coatings9050337 43

Sunil Babu Eadi, Hyeong-Sub Song, Hyun-Dong Song, Jungwoo Oh and Hi-Deok Lee
Nickel Film Deposition with Varying RF Power for the Reduction of Contact Resistance in NiSi
Reprinted from: *Coatings* 2019, *9*, 349, doi:10.3390/coatings9060349 55

Hui Liang, Xi Geng, Wenjiang Li, Adriano Panepinto, Damien Thiry, Minfang Chen and Rony Snyders
Experimental and Modeling Study of the Fabrication of Mg Nano-Sculpted Films by Magnetron Sputtering Combined with Glancing Angle Deposition
Reprinted from: *Coatings* 2019, *9*, 361, doi:10.3390/coatings9060361 65

Raúl Gago, Slawomir Prucnal, René Hübner, Frans Munnik, David Esteban-Mendoza, Ignacio Jiménez and Javier Palomares
Phase Selectivity in Cr and N Co-Doped TiO_2 Films by Modulated Sputter Growth and Post-Deposition Flash-Lamp-Annealing
Reprinted from: *Coatings* 2019, *9*, 448, doi:10.3390/coatings9070448 77

Cao Phuong Thao, Dong-Hau Kuo, Thi Tran Anh Tuan, Kim Anh Tuan, Nguyen Hoang Vu, Thach Thi Via Sa Na, Khau Van Nhut and Nguyen Van Sau
The Effect of RF Sputtering Conditions on the Physical Characteristics of Deposited GeGaN Thin Film
Reprinted from: *Coatings* 2019, *9*, 645, doi:10.3390/coatings9100645 91

Tran Anh Tuan Thi, Dong-Hau Kuo, Phuong Thao Cao, Pham Quoc-Phong, Vinh Khanh Nghi and Nguyen Phuong Lan Tran
Electrical and Structural Properties of All-Sputtered $Al/SiO_2/p$-GaN MOS Schottky Diode
Reprinted from: *Coatings* 2019, *9*, 685, doi:10.3390/coatings9100685 101

Thi Tran Anh Tuan, Dong-Hau Kuo, Phuong Thao Cao, Van Sau Nguyen, Quoc-Phong Pham, Vinh Khanh Nghi and Nguyen Phuong Lan Tran
Electrical Characterization of RF Reactive Sputtered p–Mg-In$_x$Ga$_{1-x}$N/n–Si Hetero-Junction Diodes without Using Buffer Layer
Reprinted from: *Coatings* **2019**, *9*, 699, doi:10.3390/coatings9110699 . . . . **111**

Ihar Saladukhin, Gregory Abadias, Vladimir Uglov, Sergey Zlotski, Arno Janse van Vuuren and Jacques Herman O'Connell
Structural Properties and Oxidation Resistance of ZrN/SiN$_x$, CrN/SiN$_x$ and AlN/SiN$_x$ Multilayered Films Deposited by Magnetron Sputtering Technique
Reprinted from: *Coatings* **2020**, *10*, 149, doi:10.3390/coatings10020149 . . . . **121**

About the Special Issue Editors

Alberto Palmero is a Tenured Scientist at the Institute of Materials Science of Seville, a joint center between the Spanish Council of Research (CSIC) and the University of Seville. He started his research career at the University of Seville (Spain), where he carried out his Ph.D. on the modelling and characterization of argon and oxygen plasmas employed to grow thin films by chemical and physical vapor deposition techniques (2002). After that, he continued his activity at Utrecht University (The Netherlands, 2002–2006), where he studied the plasma-assisted reactive magnetron sputtering technique and, in particular, the transport of ballistic and diffusive species in plasma gas and the associated thermal phenomena. In 2008, he got a permanent position at the Institute of Materials Science of Seville (Spain), where he led a research group aimed at fine tuning atomistic phenomena on surfaces to grow "a la carte" thin films. In 2014, he was elected Head of the "Nanostructured Functional Materials" department at the Board of the Institute of Materials Science of Seville (2014–2018). His research encompasses several areas, from plasma dynamics and plasma-surface interaction, to the description of surface nanostructuration phenomena in dynamic, far from equilibrium, situations. A key aspect of his activity is the combined approach of computer simulations and fundamental experiments to achieve accurate control of the film nanostructure, as well as the subsequent development of prototype functional devices. He has authored more than 70 publications, written 10 book chapters and has been invited to give keynote presentations at numerous international conferences and symposia. He has also authored four patents, and has won several prizes on research and innovation.

Nicolas Martin obtained a Ph.D. in Physical Chemistry from the University of Franche-Comté in 1997 and a habilitation degree (Docent) from the same university in 2005. He was a researcher at the Ecole Polytechnique Fédérale de Lausanne (Switzerland) from 1998 to 2000 in the Physics department. He got a permanent position as Associate Professor at the National Engineering School of the Ecole Nationale Supérieure de Mécanique et des Microtechniques (ENSMM) in Besançon in 2000. He became Full Professor of Materials Science in 2008. He was a visiting researcher from 2012–2013 at the University of Uppsala (Sweden), where he worked at the Angström Laboratory in the Department of Engineering Sciences, Solid State Electronics. In 2017, he spent a short sabbatical leave in the University of Alberta in Edmonton (Canada), to work in the Department of Electrical and Computer Engineering. His research is focused on the physics and technology of metallic and ceramic thin films prepared by reactive sputtering. He is also interested in nanostructuration of coatings prepared by Glancing Angle Deposition (GLAD). He was the head of the MIcro NAno MAterials & Surfaces (MINAMAS) team in the Micro Nano Sciences & Systems (MN2S) research department of the FEMTO-ST Institute in 2008 and 2009. He was previously the Deputy Director of MN2S research department from 2010 to 2014. Nicolas Martin has authored or co-authored more than 130 articles in international peer-reviewed journals, one patent, five chapters in books, one e-book, and more than 200 presentations as part of conferences, workshops and short courses.

Editorial

Advanced Strategies in Thin Films Engineering by Magnetron Sputtering

Alberto Palmero [1],* and Nicolas Martin [2],*

1. Instituto de Ciencia de Materiales de Sevilla (CSIC-Universidad de Sevilla), Americo Vespucio 49, 41092 Sevilla, Spain
2. Institut FEMTO-ST, UMR 6174, CNRS, University Bourgogne Franche-comté, 15B, Avenue des Montboucons, 25030 Besançon, France
* Correspondence: alberto.palmero@csic.es (A.P.); nicolas.martin@femto-st.fr (N.M.)

Received: 27 March 2020; Accepted: 20 April 2020; Published: 23 April 2020

Abstract: This Special Issue contains a series of reviews and papers representing some recent results and some exciting perspectives focused on advanced strategies in thin films growth, thin films engineering by magnetron sputtering and related techniques. Innovative fundamental and applied research studies are then reported, emphasizing correlations between structuration process parameters, new ideas and approaches for thin films engineering and resulting properties of as-deposited coatings.

Keywords: magnetron sputtering; nanostructures; growth mechanism; functional properties; HiPIMS; oblique angle deposition

1. Introduction

Thin films are the workhorses of many of today's innovative technologies. Entire processes, from organic electronics to aerospace to packing industries, are strongly dependent on thin films. There are many cases where a given property of thin films gave rise to an entirely new field of technology. During these last decades, thin films engineering has been changed from a laboratory curiosity to become a multi-billion euros industry worldwide. New production technologies and advanced techniques are introduced every year to add new tools to the thin film toolbox [1–4]. One of the most exciting motivations is to generate innovative thin films and original nanostructured thin films. For this purpose, recent years have witnessed the flourishing of numerous novel strategies based on the magnetron sputtering technique, aimed at the advanced engineering of thin films, such as HiPIMS, combined vacuum processes, the implementation of complex precursor gases, or the inclusion of particle guns in the reactor, among others [5–8]. At the forefront of these approaches, investigations focused on nanostructured coatings appear today as one of the priorities in many scientific and technological communities: The science behind them appears in most of the cases as a "terra incognita", fascinating both the fundamentalist, who imagines new concepts, and the experimenter, who is able to create and study new films with, as of yet, unprecedented performances [9,10].

2. Thin Films Engineering: Where Do We Stand?

Scientific and technological challenges focused on thin films engineering, along with the existence of numerous scientific issues that have yet to be clarified in classical magnetron sputtering depositions (e.g., process control and stability, nanostructuration mechanisms, connection between film morphology and properties, or upscaling procedures from the laboratory to industrial scales) have motivated us to edit a specialized volume containing the state-of-the art that put together these innovative fundamental and applied research topics.

It is systematically observed that most of the scientific and technological developments are closely linked and often limited by the performance of materials and surfaces. As a result, this last decade has seen the development of original scientific fields related to the creation of intelligent materials, functional materials, biomaterials, etc. [11,12] Structured thin films, in particular, are thus moved from laboratory curiosity to objects of high added value. They are becoming a science in themselves and complete technologies may now depend on their properties and their integration [13–15]. Various fields, such as electronics, space vehicles, decorative, etc., are highly dependent on materials and their functionality. In many cases, the scientific observation of a characteristic of a material led to the creation of a new technology. New production systems and techniques are advanced and implemented each year to create new performances in the current rush to multifunctional surfaces and materials. As a result, it became a scientific requirement to provide new opportunities for the development of components and innovative structured materials.

At the forefront of many scientific strategies, investigations focused on the surfaces and structured materials appear today as one of the priorities of many laboratories. If some groups are devoting considerable efforts to the study of nano-scaled objects, or inversely, to systems of a few tens of micrometers, the components of intermediate sizes located between the nano- and micrometer remain a "gap" of knowledge to explore. This window size appears as a "terra incognita", fascinating both for the fundamentalist, who imagines new concepts, but also for the experimenter, who is able to create and study components with unprecedented performances. It is in this dimensional window spanning the nano- to micrometer that thin films engineering strategies become more than relevant and definitely provide an extra dimension in the current race to expand the range of thin film properties.

3. This Special Issue

This Special Issue, entitled "Advanced strategies in thin films engineering by magnetron sputtering", contains five reviews and six research articles covering fundamental investigations, as well as applied research studies devoted to nanostructuration and thin films engineering produced by magnetron sputtering and related deposition methods. Without going into detail, the individual work is briefed below:

The structure, stress state and phase composition of MeN/SiN$_x$ (Me = Zr, Cr, Al) multilayered films are reviewed by Saladuhkin et al. [16] The stability of the coatings to oxidation is studied as a function of the thickness of sub-layers at the nanometric scale. The oxidation resistance of MeN/SiN$_x$ multilayers is significantly improved compared to reference monolithic films, especially by increasing the fraction of SiN$_x$ layer thickness. An optimized performance is obtained for CrN/SiN$_x$ and AlN/SiN$_x$ with nanometric periods, which remain stable up to 950 °C.

Liang et al. [17] report on the preparation of Mg nano-sculpted thin films by magnetron sputtering, implementing the glancing angle deposition technique. They demonstrate how the microstructure of the film can be tuned by adjusting deposition parameters such as the tilt angle or the sputtering pressure, which both largely influence the shadowing effect during the film deposition. They also model the growth of the material using kinetic Monte Carlo approaches, which prove the role of surface diffusion during the preparation of the film.

The paper "Gas Sensing with Nanoplasmonic Thin Films Composed of Nanoparticles (Au, Ag) dispersed in a CuO matrix" by Proença et al. presents original and interesting nano-plasmonic platforms capable of detecting the presence of gas molecules [18]. The authors show that the localized surface plasmon resonance phenomenon, LSPR, is produced by the morphological changes of the nanoparticles (size, shape, and distribution modified by thermal annealing of the films). Such an approach can be used to improve the sensitivity to the gas molecules, with the highest sensing performances for the bimetallic films.

Cougnon and Depla [19] develop thin film thermocouples as a potential way to embed sensors in composite systems, especially for their application in lightweight and smart structures. They experimentally investigate the influence of the discharge current and residual gas impurities on the Seebeck coefficient for sputtered copper and constantan thin films. These deposition parameters both lead to changes in the ratio between the impurity flux to metal flux towards the growing film. Such a parameter is assumed to be a quantitative criterion for the background residual gas incorporation in the film, and acts as a grain refiner.

The angle-resolved composition evolution of Mo-B-C thin films deposited from a Mo_2BC compound target is experimentally and theoretically investigated by Achenbach et al. [20]. The authors use TRIDYN and SIMTRA to calculate the influence of the sputtering gas on the angular distribution function of the sputtered species from the target surface, transport through the gas phase, and film composition. They show that the mass ratio between sputtering gas and sputtered species defines the scattering angle within the collision cascades in the target, as well as for the collisions in the gas phase, which influences the angle- and pressure-dependent film compositions.

The electrical and structural properties of sputter-deposited p–Mg-$In_xGa_{1-x}N$/n–Si hetero-junction diodes and Al/SiO_2/p-GaN MOS Schottky diodes are studied by Tuan et al. [21,22] Electronic transport properties by means of Hall effect measurements are comprehensively performed. Holes concentration and mobility at room temperature are determined, as well as I–V and C–V measurements at different frequencies. Other characteristics for MOS diodes are performed and compared by Cheung's and Norde's methods.

Thao et al. [23] investigate $Ge_{0.07}GaN$ films prepared by radio frequency reactive sputtering changing RF sputtering power and heating temperature conditions. Structure, optical and electrical characteristics of the films are significantly affected by both deposition parameters and with the best electronic transport properties and the lowest photoenergy produced for the deposited-150 W $Ge_{0.07}GaN$ film.

The paper "Phase Selectivity in Cr and N Co-Doped TiO_2 Films by Modulated Sputter Growth and Post-Deposition Flash-Lamp-Annealing" by Gago et al. presents how the interface engineering strategy can vary the phase occurrence in Cr and N co-doped TiO_2 (TiO_2:Cr,N) sputter-deposited films [24]. A post-deposition flash-lamp-annealing (FLA) is also used to favor anatase phase, and to give rise to dopant activation and diffusion. The authors show that using interface engineering and millisecond-range-FLA allows tailoring the structure of TiO_2-based functional materials.

In order to investigate the lowering of the contact resistance in the NiSi/Si junction, Eadi et al. systematically change the RF power implemented for the sputter-deposition of Ni thin films [25]. A post-deposition rapid thermal annealing is applied for the nickel silicide fabrication and a circular transmission line model (CTLM) procedure is developed to obtain the contact resistance. They demonstrate that Ni film resistivity can be reduced for an optimized RF sputtering power and the formed NiSi phase shows a low contact resistance.

Achour et al. [26] report on VN thin films produced by DC reactive magnetron sputtering, followed by vacuum annealing. They apply different temperatures and study the effect on the electrochemical stability and surface chemistry of the films. They particularly focus on the oxide layer formed on the VN and prove that annealing of VN films makes them an attractive candidate for long-term use in electrochemical capacitors.

In summary, this Special Issue of Coatings gathers reviews and original articles illustrating the strong potential of thin films engineering for the creation of attractive and original functional coatings based on magnetron sputtering processes. This series of publications also demonstrate the fundamental role of thin films structuration at the micro- and nanoscale for understanding growth mechanisms and generating innovative behaviors of materials and surfaces.

Funding: This research received no external funding.

Acknowledgments: We would like to warmly thank all the authors, reviewers and editors for their valuable contribution in this Special Issue of Coatings.

Conflicts of Interest: The authors declare no conflict of interest.

References

1. Gleiter, H. Nanostructured materials: Basic concepts and microstructure. *Acta Mater.* **2000**, *48*, 1–29. [CrossRef]
2. Lakhtakia, Y.; Messier, R. *Sculptured Thin Films*; SPIE Press: Bellingham, WA, USA, 2005. [CrossRef]
3. Fendler, J.H. Self-assembled nanostructured materials. *Chem. Mater.* **1996**, *8*, 1616–1624. [CrossRef]
4. Xi, J.Q.; Schubert, M.H.; Kim, J.K.; Schubert, E.F.; Chen, M.; Lin, S.Y.; Liu, W.; Smart, J.A. Optical thin-film materials with low refractive index for broadband elimination of Fresnel reflection. *Nat. Photonics* **2007**, *1*, 176–179. [CrossRef]
5. Schlom, D.G.; Chen, L.Q.; Pan, X.; Schmehl, A.; Zurbuchen, M.A. A thin film approach to engineering functionality into oxides. *J. Am. Ceram. Soc.* **2008**, *91*, 2429–2454. [CrossRef]
6. Bunker, B.B.; Rieke, P.C.; Tarasevich, B.J.; Campbell, A.A.; Fryxell, G.E.; Graff, G.L.; Song, L.; Liu, J.; Virden, J.W.; McVay, G.L. Ceramic thin-film formation on functionalized interfaces through biomimetic processing. *Science* **1994**, *264*, 48–55. [CrossRef]
7. Choy, K.L. Chemical vapour deposition of coatings. *Prog. Mater. Sci.* **2003**, *48*, 57–170. [CrossRef]
8. Hawkeye, M.M.; Brett, M.J. Glancing angle deposition: Fabrication, properties, and applications of micro-and nanostructured thin films. *J. Vac. Sci. Technol.* **2007**, *25*, 1317–1336. [CrossRef]
9. Valet, T.; Fert, A. Theory of the perpendicular magnetoresistance in magnetic multilayers. *Phys. Rev. B* **1993**, *48*, 7099–7113. [CrossRef]
10. Ohta, T.; Bostwick, A.A.; McChesney, J.L.; Seyller, T.; Horn, K.; Rotenberg, E. Interlayer interaction and electronic screening in multilayer graphene investigated with angle-resolved photoemission spectroscopy. *Phys. Rev. Lett.* **2007**, *98*, 206802–206804. [CrossRef]
11. Choi, K.; Park, S.H.; Song, Y.M.; Lee, Y.T.; Hwangbo, C.K.; Yang, H.; Lee, H.S. Nano-tailoring the surface structure for the monolithic high-performance antireflection polymer film. *Adv. Mater.* **2010**, *22*, 3713–3718. [CrossRef]
12. Spillman, W.B., Jr.; Sirkis, J.S.; Garnider, P.T. Smart materials and structures: What are they? *Smart Mater. Struct.* **1996**, *5*, 247–254. [CrossRef]
13. Arico, A.S.; Bruce, P.; Scrosati, B.; Tarascon, J.M.; van Schalkwijk, W. Nanostructured materials for advanced energy conversion and storage devices. *Nat. Mater.* **2005**, *4*, 366–377. [CrossRef]
14. Grier, D.G. A revolution in optical manipulation. *Nature* **2003**, *424*, 810–816. [CrossRef]
15. Brett, M.J.; Hawkeye, M.M. New materials at a glance. *Science* **2008**, *319*, 1192–1193. [CrossRef]
16. Saladuhkin, I.; Abadias, G.; Uglov, V.; Zlotski, S.; Janse van Vuuren, A.; Herman O'Connell, J. Structural properties and oxidation resistance of ZrN/SiN_x, CrN/SiN_x and AlN/SiN_x multilayered films deposited by magnetron sputtering technique. *Coatings* **2020**, *10*, 149. [CrossRef]
17. Liang, H.; Geng, X.; Li, W.; Panepinto, A.; Thiry, D.; Chen, M.; Snyders, R. Experimental and modeling study of the fabrication of Mg nano-sculptured films by magnetron sputtering combined with glancing angle deposition. *Coatings* **2019**, *9*, 361. [CrossRef]
18. Proença, M.; Rodrigues, M.S.; Borges, J.; Vaz, F. Gas sensing with nanoplasmonic thin films composed of nanoparticles (Au, Ag) dispersed in CuO matrix. *Coatings* **2019**, *9*, 337. [CrossRef]
19. Cougnon, F.; Depla, D. The Seebeck coefficients of sputter deposited metallic thin films: The role of process conditions. *Coatings* **2019**, *9*, 299. [CrossRef]
20. Achenbach, J.O.; Mraz, S.; Primetzhofer, D.; Schneider, J.M. Correlative experimental and theoretical investigation of the angle-resolved composition evolution of thin films sputtered from a compound Mo_2BC target. *Coatings* **2019**, *9*, 206. [CrossRef]
21. Tuan, T.T.A.; Kuo, D.H.; Cao, P.T.; Nguyen, V.S.; Pham, Q.P.; Nghi, V.K.; Tran, N.P.L. Electrical characterization of RF reactive sputtered p-Mg-$In_xGa_{1-x}N$/n-Si hetero-junction diodes without using buffer layer. *Coatings* **2019**, *9*, 699. [CrossRef]
22. Tuan, T.T.A.; Kuo, D.H.; Cao, P.T.; Pham, Q.P.; Nghi, V.K.; Tran, N.P.L. Electrical and structural properties of all-sputtered Al/SiO_2/p-GaN MOS Schottky diode. *Coatings* **2019**, *9*, 685. [CrossRef]
23. Thao, C.P.; Kuo, D.H.; Tuan, T.T.A.; Tuan, K.A.; Vu, N.H.; Na, T.T.V.S.; Nhut, K.V.; Sau, N.V. The effect of RF sputtering conditions on the physical characteristics of deposited GeGaN thin film. *Coatings* **2019**, *9*, 645. [CrossRef]

24. Gago, R.; Prucnal, S.; Hübner, R.; Munnik, F.; Esteban-Mendoza, D.; Jiménez, I.; Palomares, J. Phase selectivity in Cr and N co-doped TiO$_2$ films by modulated sputter growth and post-deposition flash-lamp-annealing. *Coatings* **2019**, *9*, 448. [CrossRef]
25. Eadi, S.B.; Song, H.S.; Song, H.D.; Oh, J.; Lee, H.D. Nickel film deposition with varying RF power for the reduction of contact resistance in NiSi. *Coatings* **2019**, *9*, 349. [CrossRef]
26. Achour, A.; Islam, M.; Ahmad, I.; Saeed, K.; Solaymani, S. Electrochemical stability enhancement in reactive magnetron sputtered VN films upon annealing treatment. *Coatings* **2019**, *9*, 72. [CrossRef]

© 2020 by the authors. Licensee MDPI, Basel, Switzerland. This article is an open access article distributed under the terms and conditions of the Creative Commons Attribution (CC BY) license (http://creativecommons.org/licenses/by/4.0/).

Article

Electrochemical Stability Enhancement in Reactive Magnetron Sputtered VN Films upon Annealing Treatment

Amine Achour [1], Mohammad Islam [2,*], Iftikhar Ahmad [2], Khalid Saeed [3] and Shahram Solaymani [4]

1. LISE Laboratory, Research Centre in Physics of Matter and Radiation (PMR), University of Namur, B-5000 Namur, Belgium; a_aminph@yahoo.fr
2. Center of Excellence for Research in Engineering Materials, Deanship of Scientific Research, King Saud University, P.O. Box 800, Riyadh 11421, Saudi Arabia; ifahmad@ksu.edu.sa
3. Department of Mechanical Engineering, King Saud University, P.O. Box 800, Riyadh 11421, Saudi Arabia; khaliduetp@gmail.com
4. Young Researchers and Elite Club, West Tehran Branch, Islamic Azad University, Tehran, Iran; shahram22s2000@yahoo.com
* Correspondence: mohammad.islam@gmail.com or miqureshi@ksu.edu.sa; Tel.: +966-54-452-3909

Received: 19 December 2018; Accepted: 23 January 2019; Published: 25 January 2019

Abstract: Vanadium nitride (VN) thin films were produced via direct-current reactive magnetron sputtering technique followed by vacuum annealing. The treatment was carried out at different temperatures for any effect on their electrochemical (EC) stability, up to 10,000 charge–discharge cycles in 0.5 M K_2SO_4 solution. The film surface chemistry was investigated by using X-ray photoelectron spectroscope (XPS) and cyclic voltammetry (CV) techniques. For the as-deposited film, the oxide layer formed on the VN surface was unstable upon K_2SO_4 immersion treatment, along with ~23% reduction in the EC capacitance. Vacuum annealing under optimized conditions, however, made the oxide layer stable with almost no capacitance loss upon cycling for up to 10,000 cycles. Annealing treatment of the VN films makes them a potential candidate for long-term use in electrochemical capacitors.

Keywords: VN films; vacuum annealing; electrochemical capacitor; XPS; cyclic voltammetry

1. Introduction

Owing to their high density and melting point, superior hardness, excellent electronic conductivity and high specific capacitance, vanadium nitride (VN) thin films offer strong potential for application in electrochemical capacitors (ECs) [1,2]. In this context, the hybrid nanostructures of VN films and carbon nanotubes have been reported to exhibit high volume capacitance, volume energy and power density [3]. Also, VN in nanocrystalline form demonstrated gravimetric capacitance of $\sim$1300 F·g^{-1}, due to successive, fast, and reversible redox reactions involving surface oxide groups and OH$^-$ ions from the electrolyte [4]. One of the major disadvantages associated with the use of VN, however, is its susceptibility to degradation that inhibits its practical application. It has been shown that cycling VN in KOH electrolyte leads to the degradation of the surface oxide layers that form at the surface of VN, and thus the capacitance decay over cycling [4]. Several factors, including film attributes such as crystallite size, morphology, surface oxide layer, etc., and EC test conditions; material loading, electrolyte concentration, and potential window, to name a few, influence the EC performance [5].

Both pure and nanocomposite VN films have been extensively explored for both structural [6,7] and functional applications [8–12]. The VN films exhibit pseudocapacitive behaviors through electric double-layer formation in the presence of OH$^-$ ions. Using N-doped carbon nanosheets/VN nanoparticles hybrid composition as the electrode, high specific capacitance with about 60% retention after 5000 cycles was reported [11]. Another study reported a reduction in areal capacitance by ~80% within first 100 cycles, when tested for 1000 cycles in 1 M KOH solution [2]. Although cycle life stability is generally assessed in KOH electrolyte solution, it is also performed in mild K_2SO_4 solution due to the relatively slow rate of decay in the EC capacitance value [13,14]. We recently reported the electrochemical properties of VN/CNT hybrid nanostructures for micro-capacitors, using potassium sulphate (K_2SO_4) electrolyte [13]. Furthermore, we suggested the preservation of the surface oxide layer for enhanced VN film stability during cycling. In this work, we demonstrate from X-ray photoelectron spectroscope (XPS) analysis that VN film annealing under certain conditions can preserve the surface oxide layer, thus enhancing the cycling life stability in 0.5 M K_2SO_4 electrolyte. Such finding have implications in the design of stable VN thin film based materials for use in ECs.

2. Experimental Procedure

Direct-current (DC) plasma reactive magnetron sputtering technique was employed to produce VN films over silicon (100) substrates. The system consisted of a magnetron sputtering gun in a stainless steel chamber in which a base pressure of $<10^{-5}$ Pa was obtained using a turbo-molecular pump. Pure argon and nitrogen with 99.99% purity were used as sputtering and reactive gases, respectively. The target was a vanadium metal of $\geq$99.9% purity. The reactive sputtering was carried out without intentional substrate heating at a pressure of 0.32 Pa. The total gas flux rate during deposition was maintained at a constant value of 40 sccm, while the flow rate for reactive N_2 gas was fixed at 35%. The power density and deposition time were kept at 12.7 W/cm^2 and 3 hr, respectively. From these deposition conditions, the VN films with an average thickness value of ~690 nm were obtained, as estimated from scanning electron microscope (SEM) examination of the film cross-sections. The as-deposited VN film is referred to as V_0.

The films were then annealed in the same chamber at different temperatures for 2 h. The pressure inside the chamber during annealing was of the order of 5.5×10^{-3} Pa. Due to low pressure during annealing, excessive oxidation of the films may be ruled out. The samples annealed at 400, 600, and 800 °C temperature were designated as V_1, V_2, and V_3, respectively.

The electrochemical measurements were performed in 0.5 M K_2SO_4 (Alfa Aesar, Ward Hill, MA, USA, 99.99%) electrolyte solution. A conventional cell with 3-electrode configuration was used in a VMP 3 multi potentiostat galvanostat (BioLogic, Seyssinet-Pariset, France) that was coupled with the EC-Lab software® V11.10.

The samples were examined under an SEM (JSM7600F; JEOL, Tokyo, Japan) by operating at 5 kV accelerating voltage and 4.5 mm working distance. The X-ray diffraction (XRD) patterns were obtained from Siemens D5000 diffractometer (Siemens, Berlin, Germany) with Bragg Brentano configuration and monochromatic CuKα1 radiation. The film surface chemistry was investigated ex situ using an X-ray photoelectron spectroscope (XPS) (Kratos Axis Ultra, Kratos Analytical Ltd, Manchester, UK). The Al Kα radiation (1486.6 eV) at 20 eV pass energy and 0.9 eV energy resolution was employed to record high-resolution spectra. As a reference, the C 1s line of 284.4 eV was used for any correction in the shift in binding energies. The XPS spectra in the V 2p core-level regions were analyzed through a peak-fitting procedure, using a Shirley background.

3. Results and Discussion

3.1. Morphology and Composition

The SEM microstructures of the as-deposited VN electrode surface and the cross-section are shown in Figure 1. As evident in Figure 1a, the film surface was observed to be comprised of nanostructured grains with a pyramid-like morphology and an average grain size of ~37 nm. Due to this specific granular morphology, the film appeared to exhibit a very high surface area, which could cause an enhancement in the specific capacitance. Also, the high film surface roughness (not measured quantitatively) implied the presence of surface pores, as evident from a few dark, depressed spots between grains. The porosity level is nevertheless very low, and is believed to be at the film surface only. Microstructural examination of the film cross-section revealed dense, columnar growth with very little porosity. From the cross-sectional microstructure (Figure 1b), the film thickness was estimated to be ~690 nm, with a corresponding growth rate of 230 nm/hr. Although not shown here, the V_1, V_2, and V_3 films did not undergo any noticeable change in surface morphology upon vacuum thermal annealing at temperatures of 400 to 800 °C.

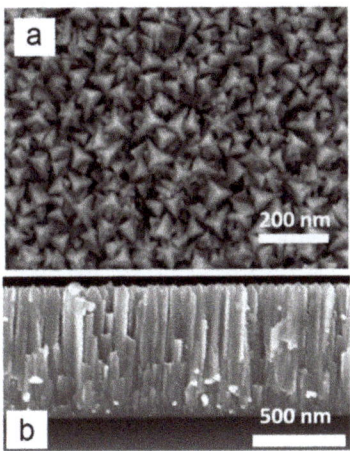

Figure 1. SEM micrographs of the as-prepared vanadium nitride (VN) film: (**a**) Surface microstructure, and (**b**) Cross-sectional view.

The XRD patterns of the as-deposited and vacuum annealed VN films are presented in Figure 2. All the samples exhibited one peak located at ~37.8° that was indexed to be (111) plane of the face-centered cubic VN (JCPDS No. 35-0768) [15]. The peak is broad, indicative of the small size of the individual crystallites in the deposited film. Upon thermal treatment, a small shift towards a greater diffraction angle was noticed, which may be attributed to stress relaxation, with consequent reduction in the lattice constant. The fcc δ-VN phase formation under the prevailing conditions of power density (12.7 W·cm^{-2}) and nitrogen gas flow (14 sccm) as well as the resulting film growth rate (3.83 nm·min^{-1}) confirm the findings reported earlier [16,17].

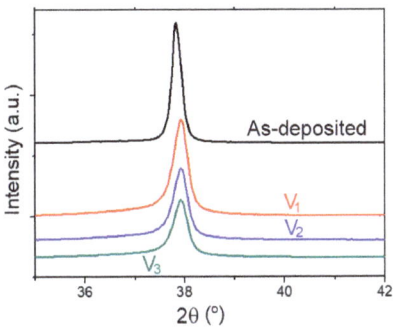

Figure 2. XRD patterns of the as-deposited and vacuum-annealed VN films.

3.2. Cyclic Voltammetry and XPS Analysis of as-Deposited VN Film

For the as-deposited VN film (V_o), cyclic voltammetry experiments were performed at a 200 mV·s^{-1} scan rate. The cyclic voltammograms of the VN films after 3 and 10,000 consecutive cycles are showcased in Figure 3. Upon repeated charge–discharge cycling, a decay in the electrochemical capacitance by ~23% was observed. It is noteworthy that the electrochemical treatment did not induce any modification in the film surface morphology or structure after cycling.

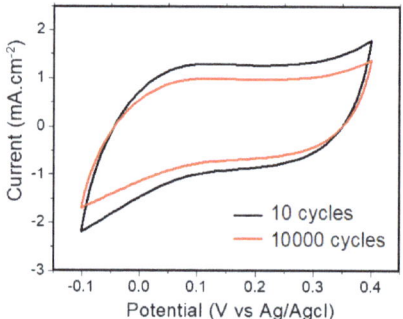

Figure 3. Cycling voltammograms of the as-deposited VN film in K$_2$SO$_4$ solution at a 200 mV·s^{-1} scan rate after 3 and 10,000 cycles.

XPS analysis was performed to determine any changes in film surface chemistry before and after EC cycling. For the as-deposited VN film (no thermal treatment), the chemical nature of the film surface is showcased in Figure 4 through V 2p and N 1s high resolution XPS spectra before and after EC cycling. The V 2p$_{3/2}$ spectral regime revealed presence of four peaks at binding energies (BE) of 513.4, 514.2, 515.5, and 517.1 eV. While the first peak is associated with VN, the latter three may be assigned to V$_2$O$_3$, VO$_2$ and V$_2$O$_5$ compositions, respectively [18,19]. It is noteworthy that, like most transition metal nitrides, surface oxidation of the VN film occurred upon air exposure and aging. It was observed that in the V 2p spectrum (Figure 4a), the intensity of the V$_2$O$_5$ peak was reduced after cycling. On the other hand, the peak intensity for VO$_x$N$_y$ remained unchanged after cycling, as revealed by high resolution spectra of the N 1s peaks (Figure 4b). The electrochemical charge–discharge cycling of the as-deposited VN film, therefore, causes deterioration of the oxide layer over the VN film surface, thus inducing a drop in the capacitance, as noticed in the cyclic voltammetry results (Figure 3).

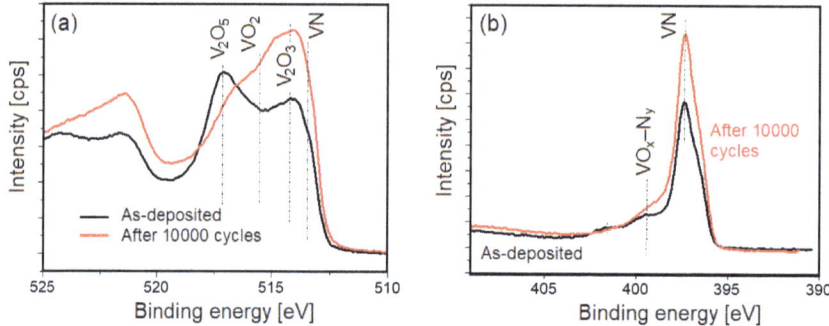

Figure 4. X-ray photoelectron spectroscope (XPS) data of the as-deposited VN film before and after 10,000 charge-discharge cycles: (**a**) Core level V 2p spectra, and (**b**) Core level N 1s spectra.

3.3. Effect of Vacuum Annealing: XPS Analysis

The high resolution XPS V 2p spectra of the as-deposited as well as thermally treated VN films are represented in Figure 5. Before the EC charge/discharge cycling, the surface chemistry of the vacuum annealed VN films was different from that of as-produced film, in the sense that 600 and 800 °C thermal treatments promoted a slightly higher degree of surface oxide layer formation. This may be attributed to the partial oxidation of the film surface at high temperatures, even under vacuum [20]. Among the vacuum-annealed VN films, only V_1 sample (annealed at 400 °C) was noticed to undergo oxide degradation after 10,000 EC cycles, as evident from comparison of the XPS spectra (Figure 5a). Quite interestingly, however, the oxide content on the VN film surfaces in the case of V_2 and V_3 films (annealed at 600 and 800 °C) did not decrease after prolonged cycling, as demonstrated in Figure 5b,c.

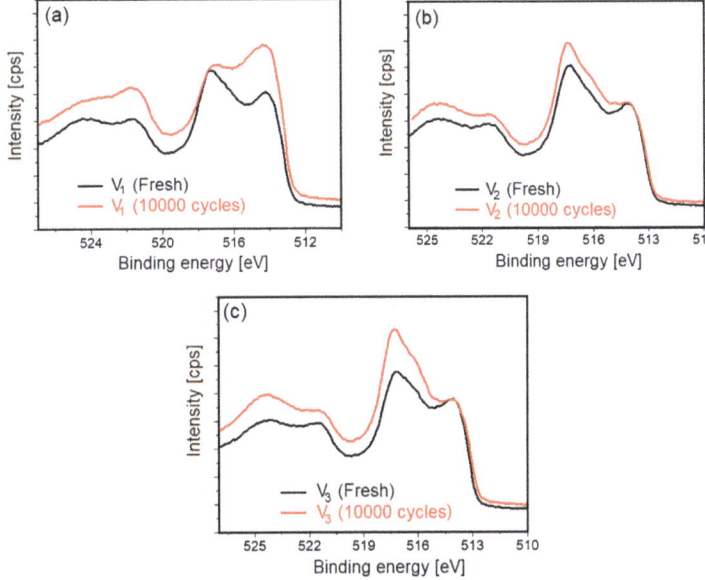

Figure 5. Comparison of the XPS core level V 2p spectra for the annealed VN films initially and after 10000 EC cycles: (a) V_1 sample, (b) V_2 sample, and (c) V_3 sample.

After EC charge/discharge tests for up to 10,000 cycles, the areal capacitance was plotted against the number of cycles, as shown in Figure 6a. For the V_2 and V_3 samples i.e., the VN films annealed at 600 and 800 °C, there was no deterioration in the areal capacitance level. Moreover, the specific capacitance was observed to be greater for the vacuum-annealed films, presumably because of an increase in the amount of oxide layer at the VN film surface. These findings indicate that vacuum annealing treatment leads to stabilization of the oxide surface layer. Factors such as formation of a thicker oxide layer, or crystallization of the oxide surface layer or both, might have been responsible for this behavior. The XPS V 2p core level spectra of the annealed VN films were recorded and deconvoluted to investigate the bonding characteristics of vanadium with oxygen and nitrogen, as presented in Figure 6b–d. The atomic percent of vanadium, bonded as V_2O_3, VO_2, and V_2O_5 compositions, are given as an inset in each case. Among the three samples examined, the presence of V_2O_5 and V_2O_3 phases was found to be maximized in case of sample V_2 (vacuum-annealed at 600 °C). Although the relative content of the VO_2 phase remained almost the same in all the films, there was an associated reduction in the VN phase content upon vacuum annealing. These observations signify the importance of annealing conditions towards any improvement in the electrochemical properties. In addition, the correlation between the capacitance improvement and the surface chemistry of the VN films may offer an insight into the influence of the V_yO_x composition on the capacitance behavior of the VN electrodes.

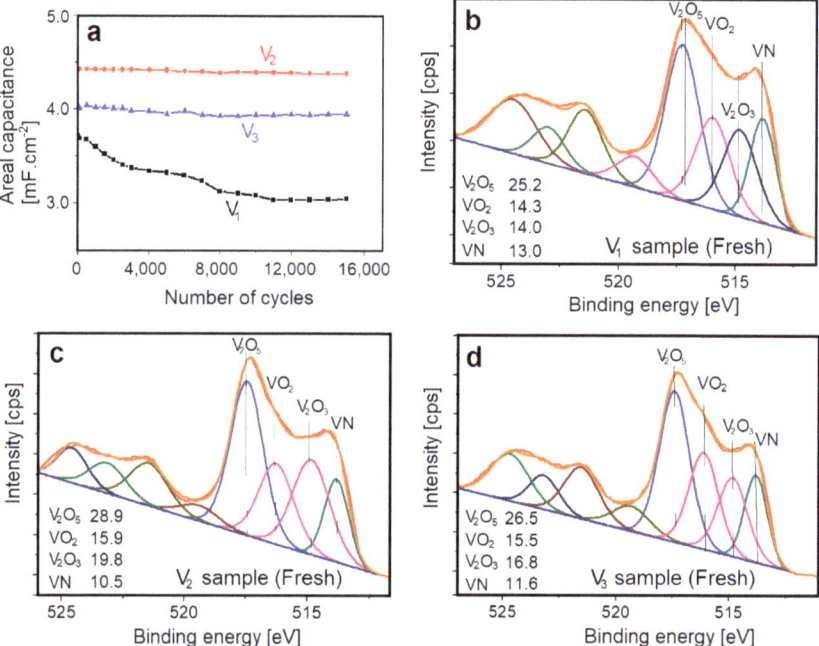

Figure 6. Cycling stability of the annealed VN film at 200 mV·s^{-1} scan rate, and (**b**–**d**) XPS V 2p core level deconvoluted spectra of the annealed VN films: (**b**) V$_1$ sample, (**c**) V$_2$ sample, and (**d**) V$_3$ sample.

4. Conclusions

Nanocrystalline VN films were produced via DC-plasma reactive magnetron sputtering followed by thermal treatment under a vacuum at 400, 600, and 800 °C. For the as-deposited and 400 °C annealed VN films, the topmost surface oxide layer was found to degrade upon treatment in K_2SO_4 electrolyte solution. This was confirmed from XPS analysis besides a ~23% capacitance loss in the case of the as-deposited VN film after 10,000 cycles. Vacuum annealing of the VN films at 600 and 800 °C led to an enhancement in electrochemical cycling stability, with an almost 100% capacitance retention, even after 10,000 cycles. Such an improvement in EC properties is speculated to originate from VN film crystallization or a thickness increase in the surface oxide layer, or both. From this finding, VN-based electrode materials may be developed with superior VN cycling stability in electrochemical energy devices.

Author Contributions: A.A. and M.I. conceived the idea and designed the experimentation scheme; A.A., K.S., and I.A. performed the experiments; M.I. and S.S. carried out data analysis; A.A. and M.I. prepared the original manuscript draft with input from other co-authors. M.I. was responsible for acquisition of funding for this research.

Funding: The authors would like to extend their sincere appreciation to the Deanship of Scientific Research at King Saud University for its funding of this research through the Research Group No. RGP-283.

Conflicts of Interest: The authors declare no conflict of interest.

References

1. Eustache, E.; Frappier, R.; Porto, R.L.; Bouhtiyya, S.; Pierson, J.F.; Brousse, T. Asymmetric electrochemical capacitor microdevice designed with vanadium nitride and nickel oxide thin film electrodes. *Electrochem. Commun.* **2013**, *28*, 104–106. [CrossRef]
2. Liu, X.; Lu, H.; He, M.; Jin, K.; Yang, G.; Ni, H.; Zhao, K. Epitaxial growth of vanadium nitride thin films by laser molecule beam epitaxy. *Mater. Lett.* **2014**, *123*, 38–40. [CrossRef]
3. Xiao, X.; Peng, X.; Jin, H.; Li, T.; Zhang, C.; Gao, B.; Hu, B.; Huo, K.; Zhou, J. Freestanding mesoporous VN/CNT hybrid electrodes for flexible all-solid-state supercapacitors. *Adv. Mater.* **2013**, *25*, 5091–5097. [CrossRef] [PubMed]
4. Choi, D.; Blomgren, G.E.; Kumta, P.N. Fast and reversible surface redox reaction in nanocrystalline vanadium nitride supercapacitors. *Adv. Mater.* **2006**, *18*, 1178–1182. [CrossRef]
5. Shu, D.; Lv, C.; Cheng, F.; He, C.; Yang, K.; Nan, J.; Long, L. Enhanced capacitance and rate capability of nanocrystalline VN as electrode materials for supercapacitors. *Int. J. Electrochem. Sci.* **2013**, *8*, 1209–1225.
6. Fu, T.; Peng, X.; Wan, C.; Lin, Z.; Chen, X.; Hu, N.; Wang, Z. Molecular dynamics simulation of plasticity in VN(001) crystals under nanoindentation with a spherical indenter. *Appl. Surf. Sci.* **2017**, *392*, 942–949. [CrossRef]
7. Sowa, M.J.; Ju, L.; Kozen, A.C.; Strandwitz, N.C.; Zeng, G.; Babuska, T.F.; Hsain, Z.; Krick, B.A. Plasma-enhanced atomic layer deposition of titanium vanadium nitride. *J. Vac. Sci. Technol. A* **2018**, *36*, 06A103. [CrossRef]
8. Yan, Y.; Li, B.; Guo, W.; Pang, H.; Xue, H. Vanadium based materials as electrode materials for high performance supercapacitors. *J. Power Sources* **2016**, *329*, 148–169. [CrossRef]
9. Zhu, L.; Li, C.; Ren, W.; Qin, M.; Xu, L. Multifunctional vanadium nitride@N-doped carbon composites for kinetically enhanced lithium–sulfur batteries. *New J. Chem.* **2018**, *42*, 5109–5116. [CrossRef]
10. Wu, Y.; Ran, F. Vanadium nitride quantum dot/nitrogen-doped microporous carbon nanofibers electrode for high-performance supercapacitors. *J. Power Sources* **2017**, *344*, 1–10. [CrossRef]
11. Tan, Y.; Liu, Y.; Tang, Z.; Wang, Z.; Kong, L.; Kang, L.; Liu, Z.; Ran, F. Concise N-doped carbon nanosheets/vanadium nitride nanoparticles materials via intercalative polymerization for supercapacitors. *Sci. Rep.* **2018**, *8*, 2915. [CrossRef] [PubMed]
12. Guo, J.; Zhang, Q.; Sun, J.; Li, C.; Zhao, J.; Zhou, Z.; He, B.; Wang, X.; Man, P.; Li, Q.; et al. Direct growth of vanadium nitride nanosheets on carbon nanotube fibers as novel negative electrodes for high-energy-density wearable fiber-shaped asymmetric supercapacitors. *J. Power Sources.* **2018**, *382*, 122–127. [CrossRef]
13. Ouldhamadouche, N.; Achour, A.; Lucio-Porto, R.; Islam, M.; Solaymani, S.; Arman, A.; Ahmadpourian, A.; Achour, H.; Le Brizoual, L.; Djouadi, M.A.; et al. Electrodes based on nano-tree-like vanadium nitride and carbon nanotubes for micro-supercapacitors. *J. Mater. Sci. Technol.* **2018**, *34*, 976–982. [CrossRef]
14. Bondarchuk, O.; Morel, A.; Bélanger, D.; Goikolea, E.; Brousse, T.; Mysyk, R. Thin films of pure vanadium nitride: Evidence for anomalous nonfaradaic capacitance. *J. Power Sources* **2016**, *324*, 439–446. [CrossRef]
15. Hajihoseini, H.; Kateb, M.; Ingvarsson, S.; Gudmundsson, J.T. Effect of substrate bias on properties of HiPIMS deposited vanadium nitride films. *Thin Solid Films* **2018**, *663*, 126–130. [CrossRef]
16. Farges, G.; Beauprez, E.; Degout, D. Preparation and characterization of V–N films deposited by reactive triode magnetron sputtering. *Surf. Coat. Technol.* **1992**, *54–55*, 115–120.
17. Hajihoseini, H.; Gudmundsson, J.T. Vanadium and vanadium nitride thin films grown by high power impulse magnetron sputtering. *J. Phys. D Appl. Phys.* **2017**, *50*, 505302. [CrossRef]
18. Sun, Y.; Liu, K.; Miao, J.; Wang, Z.; Tian, B.; Zhang, L.; Li, Q.; Fan, S.; Jiang, K. Highly sensitive surface-enhanced Raman scattering substrate made from superaligned carbon nanotubes. *Nano Lett.* **2010**, *10*, 1747–1753. [CrossRef] [PubMed]
19. Boukhalfa, S.; Evanoff, K.; Yushin, G. Atomic layer deposition of vanadium oxide on carbon nanotubes for high-power supercapacitor electrodes. *Energy Environ. Sci.* **2012**, *5*, 6872–6879. [CrossRef]
20. Glaser, A.; Surnev, S.; Netzer, F.P.; Fateh, N.; Fontalvo, G.A.; Mitterer, C. Oxidation of vanadium nitride and titanium nitride coatings. *Surf. Sci.* **2007**, *601*, 1153–1159. [CrossRef]

© 2019 by the authors. Licensee MDPI, Basel, Switzerland. This article is an open access article distributed under the terms and conditions of the Creative Commons Attribution (CC BY) license (http://creativecommons.org/licenses/by/4.0/).

Article

Correlative Experimental and Theoretical Investigation of the Angle-Resolved Composition Evolution of Thin Films Sputtered from a Compound Mo$_2$BC Target

Jan-Ole Achenbach [1,*], Stanislav Mráz [1], Daniel Primetzhofer [2] and Jochen M. Schneider [1]

1. Materials Chemistry, RWTH Aachen University, Kopernikusstr. 10, 52074 Aachen, Germany; mraz@mch.rwth-aachen.de (S.M.); schneider@mch.rwth-aachen.de (J.M.S.)
2. Department of Physics and Astronomy, Uppsala University, Lägerhyddsvägen 1, 75120 Uppsala, Sweden; daniel.primetzhofer@physics.uu.se
* Correspondence: achenbach@mch.rwth-aachen.de; Tel.: +49-241-80-25997

Received: 1 March 2019; Accepted: 20 March 2019; Published: 22 March 2019

Abstract: The angle-resolved composition evolution of Mo-B-C thin films deposited from a Mo$_2$BC compound target was investigated experimentally and theoretically. Depositions were carried out by direct current magnetron sputtering (DCMS) in a pressure range from 0.09 to 0.98 Pa in Ar and Kr. The substrates were placed at specific angles α with respect to the target normal from 0 to $\pm 67.5°$. A model based on TRIDYN and SIMTRA was used to calculate the influence of the sputtering gas on the angular distribution function of the sputtered species at the target, their transport through the gas phase, and film composition. Experimental pressure- and sputtering gas-dependent thin film chemical composition data are in good agreement with simulated angle-resolved film composition data. In Ar, the pressure-induced film composition variations at a particular α are within the error of the EDX measurements. On the contrary, an order of magnitude increase in Kr pressure results in an increase of the Mo concentration measured at $\alpha = 0°$ from 36 at.% to 43 at.%. It is shown that the mass ratio between sputtering gas and sputtered species defines the scattering angle within the collision cascades in the target, as well as for the collisions in the gas phase, which in turn defines the angle- and pressure-dependent film compositions.

Keywords: physical vapor deposition; Mo$_2$BC; Monte Carlo simulation; scattering; density functional theory

1. Introduction

Mo$_2$BC is classified as a nanolaminated material with an orthorhombic structure [1–3]. It shows a unique combination of mechanical properties, such as an elastic modulus of 470 GPa, a ratio of bulk and shear moduli of 1.73, and a positive Cauchy pressure, which are required for hard and wear-resistant coatings with moderate ductility [3,4]. Bolvardi et al. [5] successfully synthesized crystalline Mo$_2$BC at 380 °C by high power pulse magnetron sputtering (HPPMS) [6] compared to a required temperature of 580 °C during direct current magnetron sputtering (DCMS) [7]. The lower deposition temperature for the synthesis of a crystalline thin film by HPPMS was attributed to a larger adatom mobility induced by ion bombardment during HPPMS.

There are several synthesis approaches for the deposition of compound thin films, such as the utilization of reactive gases [7], co-sputtering from several targets [8,9], or targets with plugs [10]. Likewise, the employment of multi-elemental powder metallurgical composite targets is of great interest from an industrial application point of view due to the enhanced stability and repeatability [11] of non-reactive sputtering compared to reactive sputtering processes. However, it has been shown that the chemical composition of thin films deposited from multi-element targets deviates from the target composition, especially in targets with significant mass differences between their constituents, such as TiB [12–14], TiW [15–22], WB [23], SiC [24], MoSi [25,26], VC [27], NbC [28], Cr-Al-C [29,30], Ti$_2$AlC [31], Ti$_3$SiC$_2$ [32,33], and CuZnSnSe [34]. The difference in the chemical film composition was attributed to several mechanisms: (i) the mass and size differences of the target constituents and the respectively associated different angular and energy distribution functions (EDF) of the sputtered species [12,25,26,29,35]; (ii) their mean free paths, as well as the energy transfer in collisions with the sputtering gas during transport [12,13,15,21,29,35]; and (iii) different sticking coefficients and re-sputtering of the film constituents by backscattered Ar [15–20].

The compositional evolution of binary Ti-B thin films was investigated experimentally and with a Monte Carlo model based on TRIDYN (dynamic transport of ions in matter) and TRIM (transport of ions in matter) codes [12]. It was shown that the Ti/B ratio strongly depends on the gas pressure and target-substrate distance, which in a product is proportional to the number of collisions sputtered species experience within the gas phase. The model was extended to Cr-Al-C thin films–a ternary system [29].

Van Aeken et al. [36] developed a Monte Carlo code SIMTRA for the simulation of sputtered particle trajectories in a gas-phase within a definable 3D setup. Collision modelling by interatomic potentials and thermal motion of background atoms are included within the code.

From the above, it can be learned that the deviation of the chemical composition of a thin film and multi-element target can be controlled by the sputtering pressure and gas type.

Within this work, experimental data were compared to a model based on TRIDYN and SIMTRA utilized for Mo-B-C thin films to understand how the gas phase transport affects the thin film chemical composition in a system with large mass differences of the multi-element target constituents.

2. Materials and Methods

2.1. Experimental Details

Mo-B-C thin films were deposited in a high vacuum chamber assembled from a DN160 six-way cross. A base pressure of $<1.1 \times 10^{-4}$ Pa was achieved before all depositions with a combination of a rotary-vane (Edwards E2M28, Edwards, Burgess Hill, UK) and a turbomolecular pump (Pfeiffer Vacuum TPU 240, Aßlar, Germany). A self-built magnetron with Ø 90 mm was placed in the center of the chamber. A 6 mm thick Mo$_2$BC compound target (Plansee Composite Materials GmbH, Lechbruck am See, Germany) with the composition of 54.3 at.%, 24.2 at.%, and 21.5 at.% of Mo, B, and C, respectively, bonded on a Cu backing-plate, was utilized for the investigations. The target contained a major Mo$_2$BC phase with minor Mo$_2$C and MoC phases (Figure 1), as measured by a Bruker D8 Discovery general area detector diffraction system (GADDS, Bruker, Billerica, MA, USA) with Cu(Kα) radiation at 40 kV and 40 mA with a constant incident angle of $\omega = 15°$.

The thin films were deposited for 1 h onto grounded, not intentionally heated Si (100) substrates with a size of approximately 15×15 mm^2 arranged at different angular positions with respect to the target normal of $\alpha \in \{0°, \pm 22.5°, \pm 45°$ and $\pm 67.5°\}$ (Figure 2). The target-substrate distance was kept constant at 70 mm with respect to the target center point. The DC power of 100 W was applied by an ADL 1.5 kW DC power supply (ADL Analoge und Digitale Leistungselektronik GmbH, Darmstadt, Germany). The Ar and Kr pressures utilized in the depositions are summarized in Table 1.

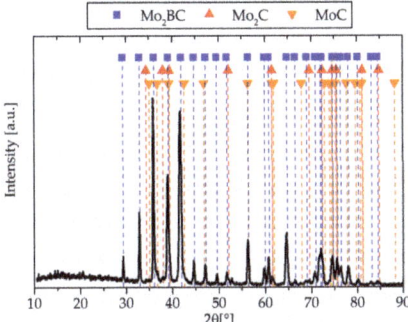

Figure 1. XRD pattern of the powder-metallurgically manufactured Mo$_2$BC compound target. Small phase fractions of Mo$_2$C and MoC were detected.

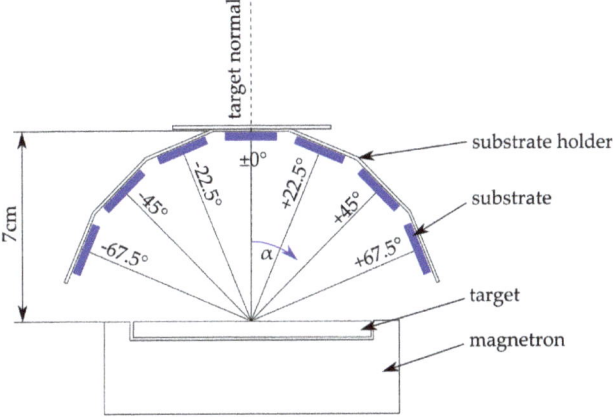

Figure 2. Experimental setup with seven substrates positioned at $\alpha \in \{0°, \pm 22.5°, \pm 45°$ and $\pm 67.5°\}$ angle arrangement with respect to the target normal and a target-substrate distance of approximately 70 mm.

Table 1. Ar and Kr gas pressures and measured target voltages which correspond to impinging ion energies for Ar$^+$ and Kr$^+$ ions.

Argon		Krypton	
Pressure (Pa)	Voltage (V)	Pressure (Pa)	Voltage (V)
0.09	401	0.09	441
0.27	344	0.26	423
0.46	328	0.45	421
0.66	324	0.64	418
0.98	314	0.96	403

The chemical composition of the deposited films was measured by energy dispersive X-ray spectroscopy (EDX) attached to a JEOL JSM-6480 scanning electron microscope (SEM, JEOL, Tokyo, Japan). The electron gun of the SEM was operated at an acceleration voltage of 5 kV. Each sample was measured 10 times. The statistical uncertainty associated with this EDX quantification of Mo, B, and C was less than or equal to 5% relative deviation. To overcome the unknown systematic uncertainty for light elements in EDX, the samples deposited at 0.66 Pa Ar with $\alpha = 0°$, $-22.5°$, $-45°$, and $-67.5°$ were quantified by time-of-flight elastic recoil detection analysis (ToF-ERDA) and used as a standard for the respective positions. The statistical uncertainty for all ToF-ERDA was <0.4% absolute. In ToF-ERDA, the relative systematic uncertainties in the specific energy loss of the constituents and primary ions of the target are assumed to range from 5% to 10%. Hence, the lower bound of the total measurement uncertainty for the EDX analysis with ToF-ERDA quantified standards ranges from 7% to 11%.

2.2. Simulation Details

The angular-resolved chemical composition of the thin films was simulated with a Monte Carlo model based on TRIDYN [37,38] and SIMTRA [36] for the sputtering process and the gas phase transport, respectively.

2.2.1. TRIDYN

The impinging ion energies of Ar^+ and Kr^+ ions in the TRIDYN simulation were set according to the experimentally measured target voltages (Table 1). To address the dependence of the surface binding energy from the surface chemistry, a matrix model was introduced [38] and modified [29] for a system containing three elements, as presented in Equation (1), where SBE_i is the surface binding energy of the i-th target element at a given target concentration c, c_i is the concentration of the i-th target element, and $SBV_{i\text{-}j}$ is the surface binding potential of the i-th and j-th elements. $SBV_{i\text{-}j}$ are assumed to be constant. Calculated angular distribution functions (ADF) and energy distribution functions (EDF) of the sputtered species are utilized in SIMTRA.

$$\begin{pmatrix} SBE_{Mo} \\ SBE_{B} \\ SBE_{C} \end{pmatrix} = \begin{pmatrix} SBV_{Mo\text{-}Mo} & SBV_{Mo\text{-}B} & SBV_{Mo\text{-}C} \\ SBV_{B\text{-}Mo} & SBV_{B\text{-}B} & SBV_{B\text{-}C} \\ SBV_{C\text{-}Mo} & SBV_{C\text{-}B} & SBV_{C\text{-}C} \end{pmatrix} \begin{pmatrix} c_{Mo} \\ c_{B} \\ c_{C} \end{pmatrix} \quad (1)$$

For the determination of the surface binding potentials, an approach based on the energy conservation law [29,38] was used and will in the following be called the energy conservation law approach. In addition, a DFT ab initio-based approach has been employed.

2.2.2. Energy Conservation Law Approach

The surface binding potential of pure elements $SBV_{i\text{-}i}$ is assumed to be equal to the enthalpy of sublimation $\Delta_{sub}H_i$. The surface binding potential of the atom pairs $SBV_{i\text{-}j}$ is calculated using Equation (2), where $\Delta_f H_{Mo_n B_m C_o}$ is the enthalpy of formation of the ternary compound and a and b are the stoichiometric factors of the elements i and j.

$$SBV_{i\text{-}j} = \frac{1}{2}(\Delta_{sub}H_i + \Delta_{sub}H_j) - \frac{1}{3}\frac{n+m+o}{2ab}\Delta_f H_{Mo_n B_m C_o} \quad (2)$$

The energy of formation per formula unit (f.u.) of $\Delta_f H_{Mo_2BC} = -1.132 \frac{eV}{f.u.}$ used in the simulations was calculated by Bolvardi et al. [4]. The enthalpies of sublimation of 6.83, 5.73, and 7.51 eV for Mo, B, and C are given in the elements.dat file of TRIDYN, respectively. In addition, enthalpies of sublimation of 6.81, 5.75, and 7.37 eV for Mo, B, and C, respectively, can be found in [39].

2.2.3. Ab Initio Approach

In addition to the TRIDYN approach, an ab initio approach based on DFT was used for the determination of the respective surface binding potentials. DFT calculations were implemented within the Vienna ab initio simulation package (VASP) [40,41]. Perdew-Burke-Ernzerhof (PBE) adjusted generalized gradient approximation (GGA) [42] was used for all calculations with projector augmented wave potential [43]. In addition, the tetrahedron method for total energy using Blöchl-corrections [44] and the reciprocal space integration using the Monkhorst-Pack scheme [45] were applied. The utilized k-point grid was $4 \times 4 \times 4$ for the (100) and (001) surfaces and $6 \times 2 \times 6$ for the (010) surface. The cut-off energy was set to 500 eV with an electronic relaxation convergence of 0.01 meV.

Considering the matrix model presented in Equation (1), the energy required to remove atoms of specific surfaces with different chemical compositions needs to be calculated. (100) and (001) surfaces, as well as different surface terminations of the (010) surface, are considered in the calculation and are illustrated in Figure 3. Subsequently, atoms are removed from the surface, creating a vacancy. The change in energy is considered to be the surface binding potential of the atom within the respective surface, as shown in Equation (3). E_i is the energy of the atom i after being removed from the surface, $E_{vac,i}^{surface,j}$ is the energy of surface j with the vacancy of atom i, and $E^{surface\,j}$ is the energy of surface j without a defect. Within DFT, the surfaces were simulated by a vacuum layer on top of the unit cell with the height of approximately 10 Å for (100) and (001) and 17 Å for (010) surfaces. Calculated SBVs for both approaches are presented in Equations (4) and (5).

$$\text{SBV}_{i\text{-}j} = E_i + E_{vac,i}^{surface,j} - E^{surface,j} \tag{3}$$

$$\text{SBV}_{\text{energy conservation law}} = \begin{pmatrix} 6.83 & 6.66 & 7.50 \\ 6.66 & 5.73 & 7.32 \\ 7.50 & 7.32 & 7.41 \end{pmatrix} \text{eV} \tag{4}$$

$$\text{SBV}_{\text{ab initio}} = \begin{pmatrix} 7.25 & 7.33 & 9.46 \\ 7.19 & 6.98 & 9.71 \\ 7.36 & 7.26 & 9.46 \end{pmatrix} \text{eV} \tag{5}$$

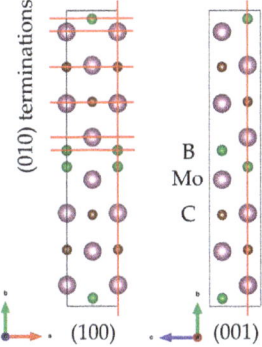

Figure 3. Considered (100), (001) surfaces and (010) surface terminations for the determination of the surface binding potentials in the ab initio approach. The colored spheres represent Mo atoms in purple, B atoms in green, and C atoms in brown. The figure was made with VESTA [46].

2.2.4. SIMTRA

Within SIMTRA simulations, 1×10^7 particles for Mo and 5×10^6 particles for B and C corresponding to a 2:1:1 target composition were transported. For the simulation setup, a cylinder with a diameter of 0.16 m and a length of 0.334 m was used. The target was positioned in the center of the simulation chamber. Seven circular substrates with a radius of 5 mm were arranged in the chamber corresponding to the actual experimental setup. The gas temperature was set to 300 K. The atomic interaction was described with the Lenz-Jensen screening function implemented in SIMTRA. Gas motion and diffusion is considered within the gas transport. The racetrack profile of the target used for the experimental work was measured by a profilometer and taken into account for the simulations. The simulations were carried out in vacuum ($p_{Ar} = 1 \times 10^{-9}$ Pa) and in Ar and Kr gaseous atmosphere at pressures utilized in the experiments (Table 1). Atoms redeposited on the target during deposition are not sputtered again within the simulation. To overcome this virtual loss of particles, atoms redeposited on the target are distributed on all surfaces within the utilized simulation chamber with respect to the initial particle distribution, including the influence of the angular distribution function. For this, the ratio of deposited atoms on a substrate divided by the total number of sputtered atoms was multiplied by the number of deposited atoms on the target surface and added to the specific substrate.

3. Results and Discussion

3.1. Experiment

The angle- and pressure-dependent film compositions for both sputtering gases, Ar and Kr, are presented in Figure 4. The target composition is indicated by black solid lines. For both sputter gases, the angle-dependence of Mo is convex, while the lighter elements B and C show a concave angle-dependence. At $\alpha \leq 22.5°$ (Figure 2), a deficiency of the heavy element (Mo) and a surplus of light elements (B and C) is measured. Mo exhibits a deficiency of up to 18 at.%, while B and C exhibit a surplus of up to 9 at.% with respect to the target composition. The opposite trend is observed for $\alpha \geq 45°$. Hence, the film composition while sputtering from a Mo_2BC target is angle-dependent, which was previously observed by Olsen et al. [35] for sputtering (metallic) alloy targets. They explained mass-dependent angular distribution functions by backscattering of light elements on the heavier elements within the collision cascade in the target [35], resulting in an enrichment of lighter elements in directions normal to the target surface. Obviously, Mo cannot be backscattered due to reflective collisions with lighter elements, such as B and C.

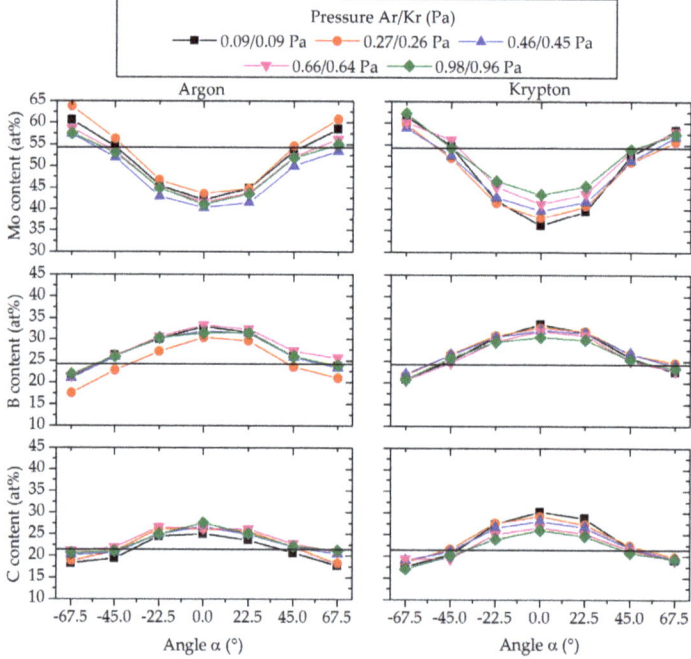

Figure 4. Angle-resolved composition evolution of the deposited thin films within the pressure range from 0.1 to 1.0 Pa. The first pressure value pertains to the Ar depositions, the second value to the Kr depositions. The average oxygen content was less than 1.5 at.% for all depositions and not considered further. The target composition is marked by the black horizontal lines.

Comparing the Mo content of Ar and Kr depositions, a clear pressure-dependence can be seen for Kr, while no significant composition changes were obtained for Ar. For Kr sputtering at $\alpha = 0°$, the Mo content changes from 36 at.% at 0.09 Pa to 43 at.% at 0.96 Pa. The chemical variation at $\alpha = \pm 45°$ is less distinct, while at $\alpha = \pm 67.5°$, the Mo content variations are within the measurement error. For gas phase scattering of B and C in Kr, the opposite trend is observed regarding the angle-dependent composition variation. However, the chemical variations due to pressure changes are within the measurement error. It is evident that an increase in pressure leads to a chemical composition closer to the nominal target composition and hence, stoichiometry. In an effort to determine the cause for the here observed sputtering gas-induced composition deviations, simulations were carried out, which allow for an independent analysis of composition deviations caused by sputtering of the target and scattering during the gas phase transport.

3.2. Simulations

The angle- and pressure-dependent film compositions with surface binding potentials (SBV's) determined by the energy conservation law and ab initio approaches, as discussed above, are presented in Figure 5 for depositions in Ar.

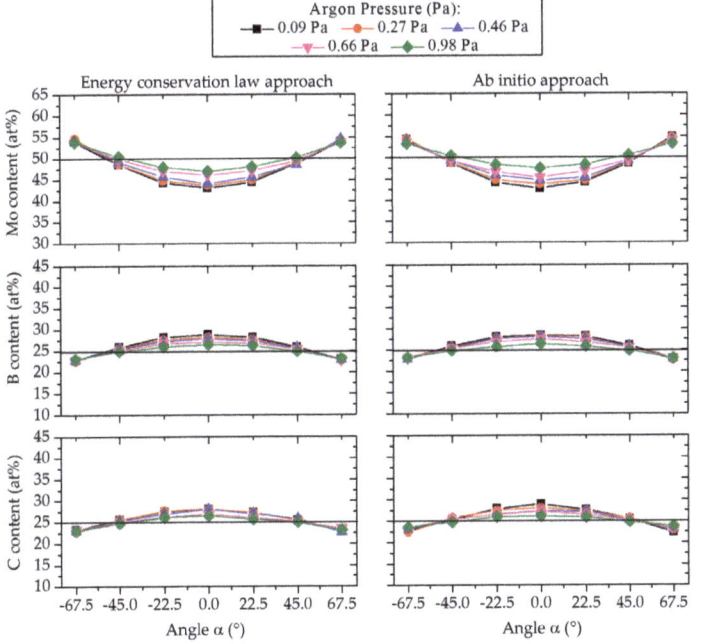

Figure 5. The simulated angle-resolved composition of thin films with the Ar pressure range from 0.09 to 0.98 Pa. Considered surface binding energies of the two approaches (**left**) energy conservation law and (**right**) ab initio. The ideal stoichiometric target composition is marked by the black horizontal lines.

The trend of the experimentally determined angle- and pressure-dependent film composition depicted in Figure 4 is reproduced. The angle-dependence of Mo is convex, while B and C show a concave angle-dependence. Films at $\alpha \leq 22.5°$ exhibit a deficiency of the heavy Mo and an enrichment of light B and C. As in the experimental data for $\alpha > 45°$, an opposite trend is observed. The maximum difference in SBVs determined by the energy conservation law and ab initio approaches is 32%. This SBV difference leads to composition differences of less or equal to 0.9 at.% and 1.1 at.% for Mo sputtered in Ar (Figure 5) and Kr (not shown), respectively. The magnitude of these composition differences cannot be resolved by EDX as the expected experimental errors are larger than the composition differences. For all simulations discussed below, SBVs determined by the ab initio approach were employed.

Pressure changes affect the target voltage and hence the ion energies impinging on the target (see Table 1). The influence of the ion energy on the ADF is illustrated in Figure 6. Within these simulations, scattering events during gas phase transport are deliberately not considered by utilizing an Ar pressure of 10^{-9} Pa. Hence, these simulations only describe sputtering, specifically the effect of the kinetic energy of Ar^+ and Kr^+ on the angle-dependent composition of the sputtered flux. These simulations will therefore be referred to as initial ADFs. Increasing the kinetic energy of Ar^+ from 314 to 401 eV (by 27%) results in absolute mean composition differences of less than or equal to 0.4 at.% for all simulations. Hence, the absolute, ion energy-induced composition changes in the sputtered flux are on average one order of magnitude smaller than the expected measurement error and hence could not be resolved by EDX measurements.

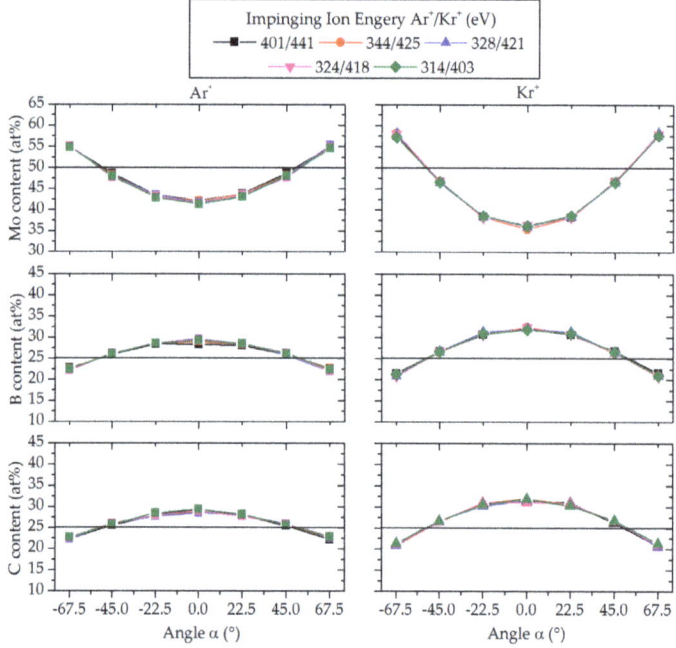

Figure 6. Angle-resolved composition evolution of the sputtered flux for different impinging ion energies of Ar^+ (**left**) and Kr^+ (**right**) ions. The first energy value pertains to Ar^+ sputtering, the second value to Kr^+ sputtering. The ideal stoichiometric target composition is marked by the black horizontal lines.

The initial ADF of Mo sputtered by Ar^+ (see Figure 6) exhibits a convex distribution, resulting in an Mo deficiency of 8 at.% at $\alpha = 0°$ with respect to a nominal Mo content of 50 at.%. At $\alpha = \pm 67.5°$, a surplus of 5 at.% Mo is obtained. Both light elements exhibit a concave distribution, resulting in a surplus of 4 at.% at $\alpha = 0°$ and a deficiency of 3 at.% at $\alpha = \pm 67.5°$ with respect to a nominal light element content of 25 at.% each. Sputtering by Kr^+ (Figure 6) leads to more pronounced convex and concave distributions for heavy and light elements, respectively. The Mo deficiency and surplus are increased to 14 at.% and 8 at.%, respectively. For both light elements, a surplus of 7 at.% and a deficiency of 4 at.% can be found at $\alpha = 0°$ and $\pm 67.5°$, respectively. Compared to Ar, the sputtering-induced differences of ADF in Kr result in larger deviations between the composition of the target and the angle-dependent sputtered flux. These results can be rationalized based on the above discussed mass-dependent reflective collisions within the target surface. In the collision cascade, only B and C can be backscattered by Mo, leading to a preferential ejection of B and C close to the target normal. Mo cannot be backscattered due to a reflective collision with lighter B or C.

Simulations of the film composition that take, in addition to sputtering at the target, the scattering events within gas phase transport into account, are shown in Figure 7. The Ar or Kr pressures are identical to the experimental pressures depicted in Table 1. Generally, the experimentally-determined angle-dependent film composition data are consistent with the simulation results.

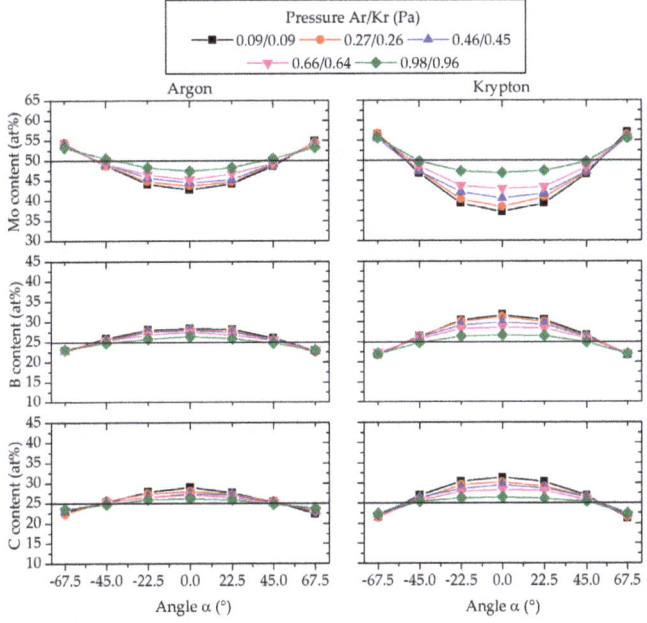

Figure 7. Angle-resolved evolution of simulated film compositions considering sputtering at the target, as well as scattering during gas phase transport. The first pressure value pertains to the Ar depositions, the second value to the Kr depositions. The ideal stoichiometric target composition is marked by the black horizontal lines.

Significant differences between the initial ADF and the ADF obtained after scattering during transport in the gas phase are obtained for Ar and Kr as the pressure is increased by one order of magnitude. An increase in Mo content at $\alpha = 0°$ of 4.7 at.% and 9.7 at.% and for both light elements a decrease of 3 at.% and 5 at.% can be obtained in Ar and Kr, respectively. At $\alpha = \pm 67.5°$, no significant pressure-induced impact on the chemical composition can be observed. Generally, the pressure-induced variations in chemical composition are more pronounced in Kr and are in good agreement with the experimentally-determined data. Comparison to the EDX composition measurement error indicates that the pressure-dependent composition variations simulated in Ar cannot be resolved experimentally.

To identify the cause of the here discussed angle- and pressure-dependent film composition variations, the angle-resolved average trajectory lengths of the sputtered species are calculated. The average trajectory length, d, is the mean distance a particle travels from sputtering at the target to deposition at the substrate surface and is maximized for scattering events at large scattering angles and short mean free paths. The pressure-dependence of d is shown in Figure 8 for Ar and Kr.

Acknowledgments: The authors gratefully acknowledge G. Dehm for fruitful discussions. The simulations were performed with computing resources granted by JARAHPC from RWTH Aachen University under Project No. JARA0131. Support by VR-RFI (contracts #821-2012-5144 & #2017-00646_9) and the Swedish Foundation for Strategic Research (SSF, contract RIF14-0053) supporting accelerator operation at Uppsala University is gratefully acknowledged.

Conflicts of Interest: The authors declare no conflict of interest.

References

1. Bovin, J.O.; O'Keeffe, M.; Stenberg, L. Planar defects in Mo_2BC. An electron microscope study. *J. Solid State Chem.* **1977**, *22*, 221–231. [CrossRef]
2. Jeitschko, W.; Nowotny, H.; Benesovsky, F. Die kristallstruktur von Mo_2BC. *Monatshefte für Chemie und verwandte Teile anderer Wissenschaften* **1963**, *94*, 565–568. [CrossRef]
3. Emmerlich, J.; Music, D.; Braun, M.; Fayek, P.; Munnik, F.; Schneider, J.M. A proposal for an unusually stiff and moderately ductile hard coating material: Mo_2BC. *J. Phys. D Appl. Phys.* **2009**, *42*, 185406. [CrossRef]
4. Bolvardi, H.; Emmerlich, J.; Music, D.; von Appen, J.; Dronskowski, R.; Schneider, J.M. Systematic study on the electronic structure and mechanical properties of X_2BC (X = Mo, Ti, V, Zr, Nb, Hf, Ta and W). *J. Phys. Condens. Matter* **2012**, *25*, 045501. [CrossRef]
5. Bolvardi, H.; Emmerlich, J.; Mráz, S.; Arndt, M.; Rudigier, H.; Schneider, J.M. Low temperature synthesis of Mo_2BC thin films. *Thin Solid Films* **2013**, *542*, 5–7. [CrossRef]
6. Sarakinos, K.; Alami, J.; Konstantinidis, S. High power pulsed magnetron sputtering: A review on scientific and engineering state of the art. *Surf. Coat. Technol.* **2010**, *204*, 1661–1684. [CrossRef]
7. Ohring, M. *Materials Science of Thin Films*, 2nd ed.; Academic Press Limited: Cambridge, MA, USA, 1992.
8. Emmerlich, J.; Högberg, H.; Sasvári, S.; Persson, P.O.; Hultman, L.; Palmquist, J.P.; Jansson, U.; Molina-Aldareguia, J.M.; Czigány, Z. Growth of Ti_3SiC_2 thin films by elemental target magnetron sputtering. *J. Appl. Phys.* **2004**, *96*, 4817–4826. [CrossRef]
9. Eklund, P.; Joelsson, T.; Ljungcrantz, H.; Wilhelmsson, O.; Czigány, Z.; Högberg, H.; Hultman, L. Microstructure and electrical properties of Ti-Si-C-Ag nanocomposite thin films. *Surf. Coat. Technol.* **2007**, *201*, 6465–6469. [CrossRef]
10. Chen, L.; Holec, D.; Du, Y.; Mayrhofer, P.H. Influence of Zr on structure, mechanical and thermal properties of Ti–Al–N. *Thin Solid Films* **2011**, *519*, 5503–5510. [CrossRef] [PubMed]
11. Eklund, P.; Beckers, M.; Jansson, U.; Högberg, H.; Hultman, L. The $M_{n+1}AX_n$ phases: Materials science and thin-film processing. *Thin Solid Films* **2010**, *518*, 1851–1878. [CrossRef]
12. Neidhardt, J.; Mráz, S.; Schneider, J.M.; Strub, E.; Bohne, W.; Liedke, B.; Moller, W.; Mitter, C. Experiment and simulation of the compositional evolution of Ti-B thin films deposited by sputtering of a compound target. *J. Appl. Phys.* **2008**, *104*, 063304. [CrossRef]
13. Mitterer, C. Borides in thin film technology. *J. Solid State Chem.* **1997**, *133*, 279–291. [CrossRef]
14. Kunc, F.; Musil, J.; Mayrhofer, P.H.; Mitterer, C. Low-stress superhard Ti-B films prepared by magnetron sputtering. *Surf. Coat. Technol.* **2003**, *174*, 744–753. [CrossRef]
15. Shaginyan, L.R.; Mišina, M.; Kadlec, S.; Jastrabik, L.; Mackova, A.; Peřina, V. Mechanism of the film composition formation during magnetron sputtering of WTi. *J. Vac. Sci. Technol. A* **2001**, *19*, 2554–2566. [CrossRef]
16. Jonsson, L.B.; Hedlund, C.; Katardjiev, I.V.; Berg, S. Compositional variations of sputter deposited Ti/W barrier layers on substrates with pronounced surface topography. *Thin Solid Films* **1999**, *348*, 227–232. [CrossRef]
17. Ramarotafika, H.; Lemperiere, G. Influence of a d.c. substrate bias on the resistivity, composition, crystallite size and microstrain of WTi and WTi-N films. *Thin Solid Films* **1995**, *266*, 267–273. [CrossRef]
18. Rogers, B.R.; Tracy, C.J.; Cale, T.S. Compositional variation in sputtered Ti-W films due to re-emission. *J. Vac. Sci. Technol. A* **1994**, *12*, 2980–2984. [CrossRef]
19. Rogers, B.R.; Cale, T.S.; Chang, Y.K. Simulation and experimental study of re-emission during sputter deposition of Ti-W films. *J. Vac. Sci. Technol. A* **1996**, *14*, 1142–1146. [CrossRef]
20. Bergstrom, D.B.; Tian, F.; Petrov, I.; Moser, J.; Greene, J.E. Origin of compositional variations in sputter-deposited Ti_xW_{1-x} diffusion barrier layers. *Appl. Phys. Lett.* **1995**, *67*, 3102–3104. [CrossRef]

21. Rossnagel, S.M.; Yang, I.; Cuomo, J.J. Compositional changes during magnetron sputtering of alloys. *Thin Solid Films* **1991**, *199*, 59–69. [CrossRef]
22. Dirks, A.G.; Wolters, R.A.M.; Nellissen, A.J.M. On the microstructure-property relationship of W-Ti-(N) diffusion barriers. *Thin Solid Films* **1990**, *193*, 201–210. [CrossRef]
23. Willer, J.; Pompl, S.; Ristow, D. Sputter-deposited WB_x films. *Thin Solid Films* **1990**, *188*, 157–163. [CrossRef]
24. Simao, R.A.; Costa, A.K.; Achete, C.A.; Camargo Jr, S.S. Magnetron sputtering SiC films investigated by AFM. *Thin Solid Films* **2000**, *377*, 490–494. [CrossRef]
25. Murakami, Y.; Shingyoji, T. Compositional difference between films and targets in sputtering of refractory metal silicides. *J. Vac. Sci. Technol. A* **1990**, *8*, 851–854. [CrossRef]
26. Yamazaki, T.; Ikeda, N.; Tawara, H.; Sato, M. Investigation of composition uniformity of $MoSi_x$ sputtering films based on measurement of angular-distribution of sputtered atoms. *Thin Solid Films* **1993**, *235*, 71–75. [CrossRef]
27. Liao, M.Y.; Gotoh, Y.; Tsuji, H.; Ishikawa, J. Deposition of vanadium carbide thin films using compound target sputtering and their field emission. *J. Vac. Sci. Technol. A* **2005**, *23*, 1379–1383. [CrossRef]
28. Liao, M.Y.; Gotoh, Y.; Tsuji, H.; Ishikawa, J. Compound-target sputtering for niobium carbide thin-film deposition. *J. Vac. Sci. Technol. B* **2004**, *22*, L24–L27. [CrossRef]
29. Mráz, S.; Emmerlich, J.; Weyand, F.; Schneider, J.M. Angle-resolved evolution of the composition of Cr-Al-C thin films deposited by sputtering of a compound target. *J. Phys. D Appl. Phys.* **2013**, *46*, 135501. [CrossRef]
30. Rueß, H.; to Baben, M.; Mráz, S.; Shang, L.; Polcik, P.; Kolozsvári, S.; Hans, M.; Primetzhofer, D.; Schneider, J.M. HPPMS deposition from composite targets: Effect of two orders of magnitude target power density changes on the composition of sputtered Cr-Al-C thin films. *Vacuum* **2017**, *145*, 285–289. [CrossRef]
31. Walter, C.; Martinez, C.; El-Raghy, T.; Schneider, J.M. Towards large area MAX phase coatings on steel. *Steel Res. Int.* **2005**, *76*, 225–228. [CrossRef]
32. Palmquist, J.P.; Jansson, U.; Seppänen, T.; Persson, P.; Birch, J.; Hultman, L.; Isberg, P. Magnetron sputtered epitaxial single-phase Ti_3SiC_2 thin films. *Appl. Phys. Lett.* **2002**, *81*, 835–837. [CrossRef]
33. Eklund, P.; Beckers, M.; Frodelius, J.; Högberg, H.; Hultman, L. Magnetron sputtering of Ti_3SiC_2 thin films from a compound target. *J. Vac. Sci. Technol. A* **2007**, *25*, 1381–1388. [CrossRef]
34. Jo, Y.H.; Mohanty, B.C.; Yeon, D.H.; Lee, S.M.; Cho, Y.S. Single elementary target-sputtered $Cu_2ZnSnSe_4$ thin film solar cells. *Sol. Energy Mater. Sol. Cells* **2015**, *132*, 136–141. [CrossRef]
35. Olson, R.R.; King, M.E.; Wehner, G.K. Mass effects on angular distribution of sputtered atoms. *J. Appl. Phys.* **1979**, *50*, 3677–3683. [CrossRef]
36. Van Aeken, K.; Mahieu, S.; Depla, D. The metal flux from a rotating cylindrical magnetron: A Monte Carlo simulation. *J. Phys. D Appl. Phys.* **2008**, *41*, 205307. [CrossRef]
37. Möller, W.; Eckstein, W.; Biersack, J.P. Tridyn-binary collision simulation of atomic collisions and dynamic composition changes in solids. *Comput. Phys. Commun.* **1988**, *51*, 355–368. [CrossRef]
38. Möller, W.; Posselt, M. TRIDYN _FZR user manual. *Qucosa* **2001**.
39. Chase, M.W., Jr.; Davies, C.A.; Downey, J.R.; Frurip, D.J.; McDonald, R.A.; Syverud, A.N. JANAF thermochemical tables, 3rd ed. *J. Phys. Chem. Ref. Data* **1985**, *14* (Suppl. 1).
40. Kresse, G.; Hafner, J. Ab initio molecular dynamics for open-shell transition metals. *Phys. Rev. B* **1993**, *48*, 13115. [CrossRef]
41. Kresse, G.; Hafner, J. Ab initio molecular-dynamics simulation of the liquid-metal–amorphous-semiconductor transition in germanium. *Phys. Rev. B* **1994**, *49*, 14251. [CrossRef]
42. Perdew, J.P.; Burke, K.; Ernzerhof, M. Generalized gradient approximation made simple. *Phys. Rev. Lett.* **1996**, *77*, 3865. [CrossRef]
43. Kresse, G.; Joubert, D. From ultrasoft pseudopotentials to the projector augmented-wave method. *Phys. Rev. B* **1999**, *59*, 1758. [CrossRef]
44. Blöchl, P.E. Projector augmented-wave method. *Phys. Rev. B* **1994**, *50*, 17953. [CrossRef]
45. Monkhorst, H.J.; Pack, J.D. Special points for Brillouin-zone integrations. *Phys. Rev. B* **1976**, *13*, 5188. [CrossRef]
46. Momma, K.; Izumi, F. VESTA 3 for three-dimensional visualization of crystal, volumetric and morphology data. *J. Appl. Crystallogr.* **2011**, *44*, 1272–1276. [CrossRef]

47. McDaniel, E.W. *Collision Phenomena in Ionized Gases*; Wiley: New York, NY, USA, 1964.
48. Behrisch, R. *Sputtering by Particle Bombardment I*; Springer: Berlin, Germany, 1981.
49. Wieser, M.E.; Berglund, M. Atomic weights of the elements 2007 (IUPAC technical report). *Pure Appl. Chem.* **2009**, *81*, 2131–2156. [CrossRef]

© 2019 by the authors. Licensee MDPI, Basel, Switzerland. This article is an open access article distributed under the terms and conditions of the Creative Commons Attribution (CC BY) license (http://creativecommons.org/licenses/by/4.0/).

Article

The Seebeck Coefficient of Sputter Deposited Metallic Thin Films: The Role of Process Conditions

Florian G. Cougnon and Diederik Depla *

Department of Solid State Sciences, Ghent University, Krijgslaan 281 (S1), 9000 Gent, Belgium; florian.cougnon@ugent.be
* Correspondence: Diederik.Depla@ugent.be

Received: 9 April 2019; Accepted: 26 April 2019; Published: 1 May 2019

Abstract: Because of their reduced dimensions and mass, thin film thermocouples are a promising candidate for embedded sensors in composite materials, especially for application in lightweight and smart structures. The sensitivity of the thin film thermocouple depends however on the process conditions during deposition. In this work, the influence of the discharge current and residual gas impurities on the Seebeck coefficient is experimentally investigated for sputter deposited copper and constantan thin films. The influence of the layer thickness on the film Seebeck coefficient is also discussed. Our observations indicate that both a decreasing discharge current or an increasing background pressure results in a growing deviation of the film Seebeck coefficient compared to its bulk value. Variations in discharge current or background pressure are linked as they both induce a variation in the ratio between the impurity flux to metal flux towards the growing film. This latter parameter is considered a quantitative measure for the background residual gas incorporation in the film and is known to act as a grain refiner. The observed results emphasize the importance of the domain size on the Seebeck coefficient of metallic thin films.

Keywords: Seebeck coefficient; background pressure; impurities; discharge current; domain size; layer thickness; sputter deposition

1. Introduction

Thin film sensing applications, such as thin film thermocouples, can be very attractive for lightweight structures, small devices or applications in need of a high temporal, or spatial resolution as they have a very low mass, reduced dimensions, and a very fast response time [1–5]. For example, thin film thermocouples are used for nanoscale thermometry [6–8], for monitoring local temperature distributions on integrated-circuits [9,10], in solid oxide fuel cells [11], and turbine engines [12], or for monitoring sudden temperature changes in cutting tools for machining explosive materials [13]. Furthermore, thin film sensors can be embedded inside composite materials without affecting the structural integrity of the material by their dimensional extent, enabling local and in situ sensing without compromise. In the case of sensing applications, typically the circuit does not draw current but an open-gate voltage is measured and linked to a physical property. Studying the behavior of the Seebeck coefficient of thin films and the relation to the deposition process is therefore an interesting field. The Seebeck effect describes the observation of an induced voltage difference when a temperature difference is applied over a metal or semiconductor. The generated voltage scales linearly to the applied temperature difference and the proportionality factor is referred to as the Seebeck coefficient S.

Thermoelectric energy is transported by mobile charge carriers [14]. Typically, metals have a high charge carrier concentration and thus a low Seebeck coefficient but a high electrical conductivity, whereas semiconductors have a low charge carrier concentration and thus a high Seebeck coefficient and low electrical conductivity. For thermoelectric applications, i.e., the conversion of heat into electrical power, it is important to combine of a low thermal conductivity (increasing temperature difference), a high Seebeck coefficient (increasing thermoelectric voltage), and a high electrical conductivity (decreasing Ohmic losses). The interplay between these three material properties is summarized under a single parameter $Z = S\sigma^2/\kappa$, called the figure-of-merit. Presently, around half of the thermoelectric research and development is focused on the maximization of the figure of merit Z for semiconductor materials such as PbTe, Bi_2Te_3 and SiGe [15]. Although semiconductors are expected to have higher Seebeck coefficients than metals, these materials are more expensive, have a more complex charge carrier transport mechanism and can quite often only be RF-sputter deposited which hinders to upscale the obtained results to industrial applications. Therefore, we solely study the behavior of the Seebeck coefficient for DC sputtered metallic thin films from a fundamental point of view. Despite the influence of layer thickness on the Seebeck coefficient of metallic films was already studied in the past [16–23], there is still much to be explored. In this study, we investigate the role of the process conditions, i.e., the discharge current and the background pressure, on the Seebeck coefficient of thin metallic films.

2. Materials and Methods

The samples discussed in the scope of this work are deposited in a cuboid stainless steel vacuum chamber with a volume of $0.7 \times 0.62 \times 0.56$ m^3. The films are grown on acetone-cleaned glass substrates (VWR International, Radnor, PA, USA) for the Seebeck measurements or either on RCA-cleaned silicon wafers in the case of film characterization. The samples were measured by X-ray diffraction (Bruker D8, Billerica, MA, USA) in a Bragg-Brentano configuration with a parallel beam bundle defined by the PolyCap. The peaks in the diffractogram were fitted to a Lorentzian-shaped curve and the Debye-Scherrer equation was used to determine the domain sizes. During all depositions, the pumping speed was kept constant ($\approx$75 l/s) and the substrate was grounded and not externally heated or cooled. The used copper and constantan sputter targets are two inch circular planar targets (purity 99.99%, Testbourne, Hampshire, UK). The films deposited for Seebeck measurements are patterned using a sputter mask specified in Figure 1. After deposition, a complementary bulk wire is attached to the thin film in order to form a conventional E-type, i.e., $Cu_{55}Ni_{45}$(film) + $Ni_{90}Cr_{10}$(wire), or T-type thermocouple, i.e., Cu(film) + $Cu_{55}Ni_{45}$(wire). The $Cu_{55}Ni_{45}$ and $Ni_{90}Cr_{10}$ wire (Goodfellow, Cambridgeshire, UK) had a diameter of 0.125 mm and a polyimide insulation. The Seebeck measurements are performed with a home-built system. The setup consists of a heating element (MeiVac substrate heater, San Jose, CA, USA) and a water-cooled copper block, separated over a distance of $\approx$ 8 cm. The bimetallic film-wire junction of the sample is centered on the heating element, whereas the electrodes are centered on the copper block. Thermal paste (Dow Corning, Midland, MI, USA) was applied at both ends of the glass substrate in order to enhance the heating or cooling transfer. The water-cooled copper block is kept at a constant temperature of 11 °C. The heating element is PID-controlled and ramps from room temperature to 60 °C at a rate of 1 °C/min. Due to the imposed temperature difference, an open-circuit thermoelectric voltage arises at the electrodes of the film-wire thermocouple. The output of the thermocouples was logged by means of a TC01 (National Instruments, Austin, TX, USA), a thermocouple measurement device with built-in software for data acquisition. In order to provide a good electrical contact with the film for the read-out, a fine copper brush was clamped onto the thin film. The hot junction and cold junction temperature were measured by means of a K-type thermocouple. A Labview code managed all the data and plotted the measured voltage as function of the temperature difference between the hot and cold junction. A straight line was fitted to the data ($\approx$200 measurement points) and the slope of this line was used to determine the Seebeck coefficient. The experimental values for the E- and T-type bulk thermocouples (i.e., wire + wire) are used as reference and are (67.3 $\pm$ 1.4) μV/°C and (46.3 $\pm$ 0.8) μV/°C respectively.

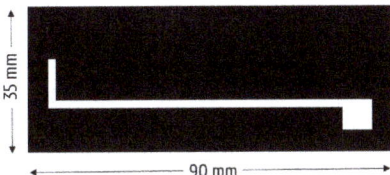

Figure 1. Design of the sputter mask used for pattering the thermocouple leg. The mask is fabricated out of stainless steel and has a thickness of 0.5 mm.

The influence of contamination on the thin film Seebeck coefficient was investigated by progressively increasing the background pressure before deposition. This was achieved by leaking atmospheric air into the vacuum chamber with a mass flow controller (MKS, Andover, MA, USA). After the background pressure is stabilized at a desired value, the sputter gas pressure is set by the introduction of argon gas (Argon 5.0 Praxair, Danbury, CT, USA) in the system. Finally the magnetron discharge is ignited and the target is sputtered at constant current (0.6 A). The power towards the magnetron is delivered by a Hüttinger 1500DC power supply (Ditzingen, Germany). The target-to-substrate distance and the argon pressure were kept constant during all depositions at 10 cm and 0.6 Pa. Except otherwise specified, all films, irrespective of the material, had a thickness of 350 ± 30 nm. The impurity-to-metal impingement flux ratio τ is used as a measure for the degree of contamination. This parameter is defined as the ratio between the impurity flux and the material flux impinging on the substrate surface, i.e., $\tau = F_i/F_m$. The material flux was calculated from the measured thickness (Taylor-Hobson profilometer, Leicester, UK), the deposition time, and the film density as measured by x-ray reflectometry (Bruker D8). The impurity flux was calculated based on the Maxwell-Boltzmann distribution as presented in previous work [24].

The depositions for the experiment on the influence of the discharge current were performed under the same process conditions as the samples discussed above, i.e., target-to-substrate distance 10 cm and argon pressure 0.6 Pa. The experiment was performed at two different background pressures. One series was deposited at a low background pressure (3.50×10^{-4} Pa) and one series at a high background pressure (7.50×10^{-3} Pa). The magnetron was sputtered in current-controlled way and the discharge current was varied from 0.6 to 0.1 A. This was done in a random way in order to exclude any effects related to target erosion and/or chamber heating.

3. Results

3.1. Impurities

Figure 2 presents the results for the deviation of the thin film Seebeck coefficient from the bulk value, i.e., $|\Delta S| = |S_B - S_F|$, plotted as function of the impurity-to-metal impingement flux ratio τ for both copper (a) and constantan (b). In the low-impurity regime, i.e., $\tau \ll 1$, the Seebeck coefficient is not significantly affected by an increased background pressure. The average deviation from the bulk value in this low-impurity regime is represented by the dotted line for both copper (1.9 μV/°C) and constantan (2.61 μV/°C). In the high-impurity regime, i.e., $\tau \gg 1$, the Seebeck coefficient is strongly affected by an increasing degree of contamination. For both figures, error bars are included on a single marker but are valid for all data points in the respective plots. In order to simplify the rationale in Section 4.3, the striped lines in the figures represent the relation $|\Delta S| \sim \tau^{1/2}$ where the proportionality factor is fitted to coincide with the data in this high-impurity regime. This specific description of $|\Delta S|$ as function of the impurity-to-metal ratio τ embodies the thought of a relation between $|\Delta S|$ and the domain size D, more specifically $1/D$. The idea for this relation originates from a previous work [24], where we elaborated a model that accounts for the experimental observation $D \sim \tau^{-1/2}$ in the high-impurity regime. The agreement between the experiment and the proposed relation $|\Delta S| \sim \tau^{1/2}$

is evaluated by means of a chi-squared test. For copper and constantan, we find a value $\chi^2 = 1.98$ (12 data points) and $\chi^2 = 3.21$ (9 data points) respectively. In both cases, the significance of the description is guaranteed within the 95% confidence interval. The full gray lines in Figure 2 represent the maximal deviation $|\Delta S|_{max}$ which is found when $S_F \rightarrow 0$ and thus $|\Delta S|_{max} = |S_B|$. We find for copper and constantan a value of 6.2 µV/°C and 40.1 µV/°C [25] respectively. Copper has a much smaller Seebeck coefficient in comparison to constantan. This hinders the electronic measurement and results in a high signal-to-noise ratio. As the observed trends are similar for both materials, we focus our research on constantan as the latter material has a higher Seebeck coefficient.

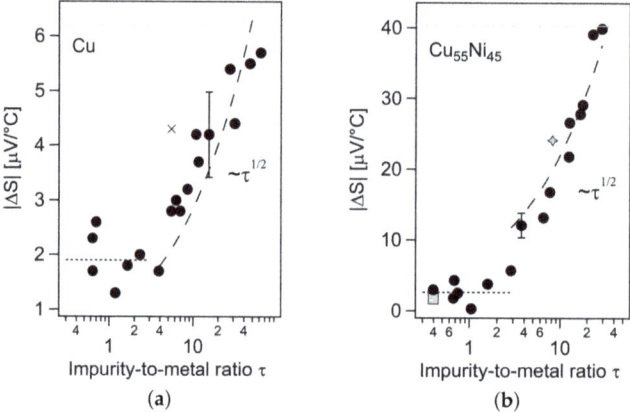

Figure 2. The deviation of the thin film Seebeck coefficient from the bulk value plotted as function of the impurity-to-metal impingement flux ratio for (**a**) copper and (**b**) constantan. The depositions are performed at a discharge current of 0.6 A, an argon pressure of 0.6 Pa and a target-to-substrate distance of 10 cm. The impurity-to-metal ratio was increased by leaking atmospheric impurities inside the vacuum chamber. The full gray lines in the figures indicate the maximal deviation $|\Delta S|_{max}$ for copper (6.2 µV/°C) and constantan (40.1 µV/°C). The dotted lines represent the average deviation from the bulk value in the low-impurity regime for copper (1.9 µV/°C) and constantan (2.61 µV/°C) and the striped lines represent the power law $|\Delta S| \sim \tau^{1/2}$. Error bars are included on a single marker but are valid for all data points in the respective plots. The data point represented by the cross marker (**a**) was excluded from the fit for the proportionality factor. The gray square and diamond markers (**b**) represent data points from the experiment discussed in Section 3.2.

3.2. Discharge Current

The results for the influence of the discharge current on the thin film Seebeck coefficient are presented in Figure 3. The square markers and diamond markers represent depositions performed at a background pressure of 3.50×10^{-4} Pa and 7.50×10^{-3} Pa respectively. The values for $|\Delta S|$ are significantly larger for the series deposited at a high background pressure compared to the series deposited at a low background pressure, an observation in accordance to the results discussed in Figure 2. As an illustration, the data points obtained in this experiment which corresponds to the same deposition conditions as the previous experiment, i.e., a discharge current of 0.6 A, are represented by the gray (square and diamond) markers on the plot in Figure 2. For both series, the measurements indicate an increasing deviation of the film Seebeck coefficient from the bulk value with decreasing discharge current.

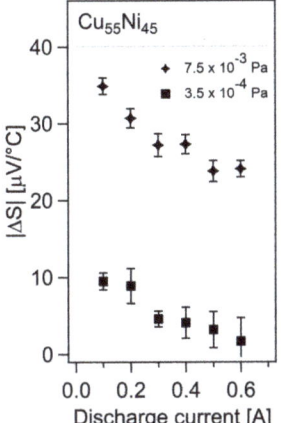

Figure 3. The deviation of the thin film Seebeck coefficient from the bulk value measured as function of the discharge current for 2 different background pressures: 3.50×10^{-4} Pa (square markers) and 7.50×10^{-3} Pa (diamond markers). The argon pressure was set to 0.6 Pa and the target-to-substrate distance was 10 cm. The full gray line indicates the maximal deviation $|\Delta S|_{max}$.

3.3. Domain Size

The domain sizes corresponding to the series deposited at varying discharge current at both low and high background pressure are plotted in Figure 4a. The domain size decreases with increasing background pressure and decreasing discharge current. As the discharge current is varied from 0.6 A up to 0.1 A, the deposition speed was varied from 1.28 nm/s up to 0.23 nm/s respectively. Given the constant background pressure used within each series, the ratio of the impurity-to-metal impingement flux ratio τ thus increases with decreasing discharge current. The variation in background pressure between both series did not have any influence on the deposition speed. Figure 4b is based on the same data for the domain size as in Figure 4a but presented in a normalized way and as function of the impurity-to-metal impingement flux ratio τ. The data for the domain size have been normalized in order to allow a better comparison with the additional data set (gray circles) included in Figure 4b which originates from a previous experiment [24]. The uppermost dotted line in the figures indicate the averaged domain size in this low-impurity regime. The domain size in the high-impurity regime ($\tau \gg 1$) is strongly refined by the presence of the impurities during growth. Based on previous research [24], we expect a relation $D \sim \tau^{-1/2}$ in this regime. This relation is represented by the striped line and further consolidated by the gray markers taken from the previous work. In the low-impurity regime ($\tau \ll 1$), the domain sizes are larger and less affected by an increase in impurity-to-metal impingement flux ratio. It must be noted here that in the determination of the domain sizes from the XRD data, we did not account for microstrain contributions. However, this analysis was carried out on the additional data set (gray circles Figure 4) from previous work, but no systematic variation in microstrain was observed.

Figure 5 summarizes the effect of background pressure (square markers vs. diamond markers) and discharge current (square markers and diamond markers) on $|\Delta S|$ described as function of the inverse of the domain size, i.e., $1/D$. Additional data from the work of Barber et al. [26] are also included on the figure (round gray markers and unfilled blue markers). Whereas the data in the latter work are presented as $|\Delta S|$ as function of the FWHM, they are recalculated here using the Scherrer-equation in order to present them as function of $1/D$.

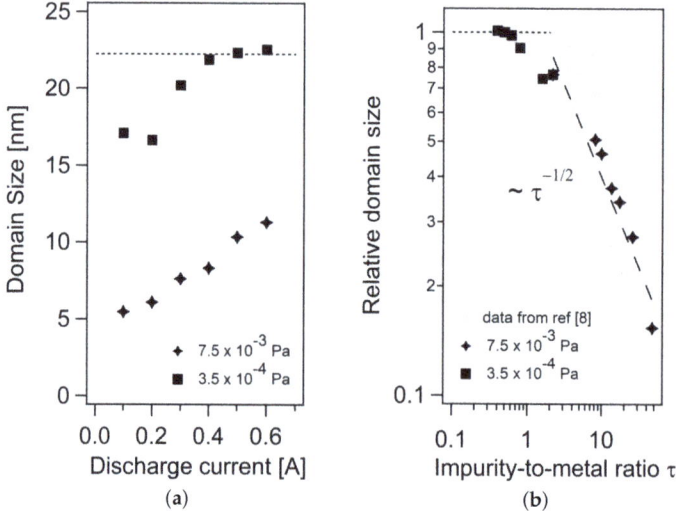

Figure 4. (a) Domain size as function of the discharge current; (b) Domain size presented as function of the impurity-to-metal impingement flux ratio. The square markers represent the deposition performed at low background pressure (3.50×10^{-4} Pa), the diamond markers represents the depositions performed at high background pressure (7.50×10^{-3} Pa). The dotted lines in the figures represents the average domain size in the low-impurity regime. The striped lines (b) represent the relation $D \sim \tau^{-1/2}$.

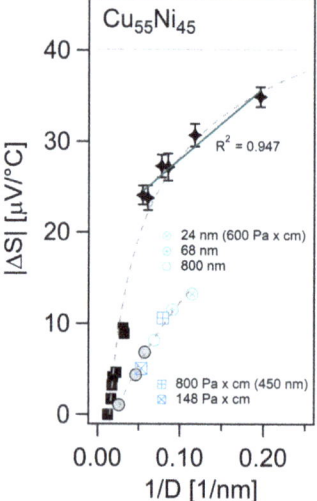

Figure 5. Deviation of the thin film Seebeck coefficient from the bulk value as function of the inverse of the domain size. The data derived from the work of Barber et al. [26] on sputter deposited constantan films are represented by the blue square unfilled markers (varying $P \times d$, constant thickness), the blue round unfilled markers (varying thickness, constant $P \times d$) and the round gray filled markers (both varying $P \times d$ and layer thickness). The full gray line represents the value for the maximal deviation $|\Delta S|_{max}$. The gray dotted lines are a guide to the eye.

4. Discussion

The work of Barber et al. [26] investigated the influence of the layer thickness and $P \times d$, i.e., the product of the sputter gas pressure P times the target-to-substrate distance d, on the Seebeck coefficient of sputter deposited constantan films. The experiments showed that for a fixed $P \times d$-value, there is an increasing deviation $|\Delta S|$ from the bulk value as the film grows thinner. A similar observation was done by Gierczak et al. [27]. This is a classic observation of the size-effect. Inversely, at fixed layer thickness, $|\Delta S|$ increases with increasing $P \times d$-values. Only the data for which their respective domain size D could be calculated are included in the plot in Figure 5. The influence of the layer thickness on the film Seebeck coefficient at fixed $P \times d$ is presented by the unfilled blue round markers, the influence of $P \times d$ at constant layer thickness is represented by the unfilled blue square markers. The round filled gray markers in Figure 5 represent combinations of a varying thickness together with a unspecified $P \times d$-value (thickness range 2000–90 nm).

The main conclusion from the work of Barber is the direct correlation between the microstructure and the thin film Seebeck coefficient. The effect of the layer thickness was attributed to the morphological development as the film thickens. The dependency of the Seebeck coefficient on $P \times d$ is explained by thermalization effects affecting the microstructure. The role of the layer thickness is further discussed in Section 4.1, whereas the hypothesis of thermalization is confronted with our observations in Section 4.2.

4.1. Role of Layer Thickness

The electrical properties of a thin metal film deviate from the bulk as soon as one of the dimensions of the metal is restricted in the order of the electron mean free path length. The mean free path length of an electron in a metal is typically of the order of tens of nanometers [28], and is therefore easily restricted by the spatial dimensions of a thin film. The effects caused by this reduced mean free path are typically called "size-effects". For example, this is a major challenge in the field of microelectronics for contacting and connection paths in the sub-50 nm range. Over time, many models have been proposed in order to describe and understand the origin of these effects. The Fuchs-Sondheimer model [29] describes the increase in thin film resistivity from the perspective of an enhanced electron scattering caused by the decreased layer thickness. The model is described as function of l_0/t, i.e., the ratio of the electron mean free path l_0 to the layer thickness t, and a parameter p, representing the fraction of specular reflected electrons at the external film surfaces (e.g., substrate-film and film-air). The thinner the film, the more electron scattering events and thus the higher the electrical resistivity of the film, i.e., $\rho \sim 1/t$. Later on, a similar model was proposed by Mayer [30] in order to describe the thermoelectric film properties. Analogously to the Fuchs and Sondheimer description for the electrical film resistivity, Mayer predicts the deviation of the thin film Seebeck coefficient from the bulk value to scale with the inverse of the layer thickness, i.e., $|S_B - S_F| = |\Delta S_F| \sim l_0/t$. Thus according to the Mayer theory, ΔS_F plotted as function of $1/t$ yields a straight line. This behavior can indeed be verified by an abundant amount of experimental data present in the literature. For example, this is reported for sputtered $Ni_{90}Cr_{10}$ [31], for evaporated Cu [4,18,32,33], Ag [18,34,35], Au [35], Al [36], Fe [3], Sn [21], Sb [37], Bi films [37–39], or electron-beam evaporated Ni [40].

Sputtered thin films grow in a columnar fashion with columns growing in a lateral direction with respect to the substrate and with a mean diameter D_{in}, referred here to as the in-plane domain size [41,42]. In most cases, the in-plane domain size is measured by means of characterization techniques such as TEM, SEM, AFM or EBSD. When a sample is measured by means of XRD, one measure the degree of crystallographic coherence in the out-of-plane direction. The more (X-ray) interference events at consecutive crystallographic planes through the thickness of the sample, the sharper the peakwidths (FWHM) of the resulting diffractogram. Using the Scherrer-equation ($D_{out} \sim (FWHM)^{-1}$), sharp peaks with a small FWHM yield large (out-of-plane) domains D_{out}. For sputtered as-deposited films, the in-plane domain size D_{in} measured by TEM is very similar to the out-of-plane domain size D_{out} as measured by XRD [43]. Therefore, in what follows, we will use the out-of-plane domain size as measured by XRD as a representative measure for the in-plane domain size and both in-plane and out-of-plane are simply referred to as the domain size D.

For thin films grown by physical vapor deposition in general, it is an observation that the domain size increases with the film thickness, i.e., $D \sim t^\chi$, with χ a value between 1/3 and 1/2 [44–48], depending on the growth mode, substrate temperature and more. This can be verified with experimental data presented in the literature for sputtered Cu [49–51], Ag [52,53], Mo [42], Fe [54], FeCo [55], CuNi [56], NiCr [56] and Sb_2Te_3 [57], or for evaporated $Ni_{80}Fe_{20}$ (permalloy) films [58] or ion beam-evaporated Ni film [40]. Variations in film thickness are thus linked with variations in the domain size. The thinner the film, the smaller the resulting domains. Typically, for sputtered as-deposited thin films, $D < t$ [42,51–57]. Therefore, the size-depend effects observed for films with decreasing thickness can thus either be attributed to variations in D or t [59]. Indeed, when electrons flow through a thin film under influence of an electric field (resistivity) or temperature difference (thermoelectricity), the net flow is parallel to the substrate and film surface and thus perpendicular to the growth direction (out-of-plane) of the grains. This, in combination with the observations that $D < t$, makes it very likely that a much higher contribution of electron scattering at grain boundaries is to be expected for sputtered as-deposited films, rather than the scattering contribution from the external film surfaces as predicted by the Fuchs-Sondheimer or Mayer theory. The idea of an additional grain-boundary scattering contribution to the Fuchs-Sondheimer scattering model was already implemented under the Mayadas-Shatzkes model [41] for the description of the thin film resistivity. The shortcomings of the Fuchs-Sondheimer model to describe experimental data and the many experimental validations of the MS-model proved the dominant character of grain-boundary scattering [41,50,60,61]. Later on, Deschacht et al. [62] elaborated a similar analytical model for the description of the thermoelectric power of polycrystalline semimetal films taking into account the effects of grain boundaries. The strong contribution grain-boundary scattering could have on the Seebeck coefficient was also already hinted by other authors [26,40,63]. However, to our knowledge, elaborated experimental studies on the Seebeck coefficient of sputtered thin films, their relation to deposition conditions [64] and proof for the dominant character of the domain size on the Seebeck coefficient of (metallic) thin films [26] remains very limited. Based on the above discussion and in agreement with the latter work of Barber et al. [26], we conclude that the microstructural development during film growth can account for the observed variations in the film Seebeck coefficient as function of the layer thickness. In what follows, the layer thickness is kept constant (350 nm) for all depositions in order to exclude this influence from the discussion.

4.2. Process Parameters

The main effect of varying deposition conditions such as the discharge current I or the argon pressure P and the target-to-substrate distance d, is a variation in energy flux and/or material flux arriving on the substrate. As both the energy flux and the material flux are system-dependent parameters, i.e., not easily transferable towards other vacuum systems, we prefer to project them on system-objective parameters such as the energy per arriving adatom (EPA) and the impurity-to-metal impingement flux ratio τ. The EPA has different energetic contributions such as a contribution from the kinetic energy of the sputtered particles or reflected neutrals, from electronic contributions, and from plasma radiation. Whereas the latter two contributions are mainly affected by the target-to-substrate distance d, the former two are mainly affected by $P \times d$, i.e., the product of the argon pressure and the target-to-substrate distance. As discussed in the work of Z. Barber et al. [26], $P \times d$ is a measure for the degree of thermalization. As thermalized adatoms have a restricted mobility on the substrate surface, large $P \times d$-values result in smaller grains.

In contrast to $P \times d$, the EPA remains mainly indifferent under a variation in discharge current. This because when the discharge current is decreased, both the decrease in material flux arriving on the substrate as well as the total power dissipated in the system approximately scale in a linear way [56]. The EPA is thus not significantly affected by a variation in the discharge current. The experimental result presented in Figure 3 is therefore not compatible with the idea of a variation in film Seebeck coefficient caused by thermalization effects. However, as the deposition speed drops with a decreasing discharge current, this causes the impurity-to-metal ratio τ to increase. It is shown in Figure 4 (right) that this latter effect is responsible for the decrease of the domain size with decreasing discharge current. Analogously, the impurity-to-metal ratio τ increases with increasing $P \times d$ as $P \times d$, and especially d, affects the deposition speed [56]. It can also be verified in the literature that for an increasing value for $P \times d$, the domain size decreases [65–67]. Our observations therefore indicate that a more consistent description for the effect of the discharge current on the thin film Seebeck coefficient (as presented in Figure 3) can be given from the perspective of refined grains - in coherence with the results reported in the work of Barber - but caused by an increased impurity-to-metal impingement flux τ rather than caused by thermalization effects. Furthermore, the $P \times d$-values used in the latter work, i.e., in the range of 150–800 Pa mm, are very high in comparison to the $P \times d$-value used in this work, i.e., 60 Pa mm. The EPA values in the work of Barber are thus expected to be much lower than in this work. According to SiMTra simulations, i.e., a kinetic Monte Carlo code for the simulation of metal transport through the gas phase [68], the energy of the sputtered particles in the range of 150–800 Pa mm is estimated betzeen 1.5 eV and $3/2kT$ (T at 300 K) respectively and around 6 eV for the $P \times d$-value (60 Pa mm) used in this work. The simulations indicate fully thermalized sputtered particles, i.e., $E = 3/2kT$, at a $P \times d$-value close to 400 Pa mm. Therefore, it seems less likely that the variations in $|\Delta S|$ as function of $P \times d$ (see Figure 2 in [26]) could be attributed to thermalization effects, especially for the data discussed where $P \times d > 400$ Pa mm. Of course, other contributions could still contribute to the energy balance for the adatoms, although for this material-sputter gas combination, the amount of reflected neutrals is expected to be low [69,70] and the electronic and radiative contributions rather scale as function of d and not as $P \times d$ which is the expected dependency based on scattering events for massive particles. However, in order to demonstrate our point-of-view with respect to the importance of impurity incorporation, in the following section, both the layer thickness and the EPA are kept constant to eliminate thermalization effects and solely the impurity-to-metal ratio τ is altered.

4.3. Impurities

As already discussed in previous work [24], atmospheric gas impurities present in the vacuum chamber during film growth can act as grain refiners. In general, in the high-impurity regime ($\tau \gg 1$), the average domain size D decreases in a material-independent way with an increasing impurity-to-metal impingement flux ratio τ according to $D \sim \tau^{-1/2}$ [24]. This effect was experimentally verified in Figure 4 (right). The relation embodies the thought of a nucleation-dominated growth model. When the spatial distribution of impinging impurities on the surface is of the order of the characteristic length of the diffussing adatoms, the impurities act as active nucleation centra and restrict the adatom diffusion with refined grains as a consequence. The results presented in Figure 2 showed how the film Seebeck coefficient was strongly affected in this high-impurity regime. The similarity in behavior between the domain size $D \sim \tau^{-1/2}$ and the Seebeck coefficient as function of the impurity-to-metal ratio $|\Delta S| \sim \tau^{1/2}$ (see Figure 2) further empowers the perspective to describe the Seebeck coefficient as a film property determined by the domain size. Based on our assumptions, an approximate linear relationship $|\Delta S| \sim 1/D$ can thus be expected in this high-impurity regime. As presented in Figure 5 (diamond markers), this prediction can be fairly well validated. Furthermore, in analogy to the work of Liu et al. [71] where the deviation in film Seebeck coefficient due to the size-effect is employed for manufacturing single-metal thermocouples, the observation of a decreased thin film Seebeck coefficient as function of τ could be exploited for creating thermocouples where both legs consist of the same material but are deposited at a different τ-value.

5. Conclusions

The effect of impurities, layer thickness and deposition conditions such as discharge current and $P \times d$ on the thin film Seebeck coefficient is discussed. It was emphasized here how a variation in each of these deposition variables can affect the domain size. In view of the diversity of experimental variations which could be summarized as function of the domain size, more specifically, $1/D$, it is justified to conclude that rather than the layer thickness, the domain size has a key role in the deviation of the thin film Seebeck coefficient from the bulk value. Therefore, the influence of deposition conditions on the Seebeck coefficient can be evaluated by understanding their effect on the domain size. This conclusion is mainly in coherence with a the work on sputter-deposited constantan by Barber et al. [26], i.e., there is a direct correlation between the microstructure and the Seebeck coefficient, but where our results indicate that the origin of the grain refinement can be attributed to an increased impurity incorporation during growth rather than to thermalization effects. Of course, it is not straightforward to discriminate between the active mechanisms as an increase in the degree of thermalization and a decrease in the deposition rate, and thus an increase in impurity-to-metal impingement flux ratio τ, are unambiguously linked through an increasing target-to-substrate distance d. Although the data of the experiments performed in this work and in the work of Barber are quantitatively shifted in $|\Delta S|$—which we believe can be attributed to experimental differences—this does not compromise on the generality of the discussion as the qualitative response of the Seebeck coefficient to the variations in domain size is very comparable.

Author Contributions: Conceptualization, F.G.C. and D.D.; methodology, F.G.C.; software, F.G.C.; validation, F.G.C.; formal analysis, F.G.C. and D.D.; investigation, F.G.C.; resources, F.G.C.; data curation, F.G.C.; writing–original draft preparation, F.G.C.; writing–review and editing, D.D.; visualization, F.G.C.; supervision, D.D.; project administration, D.D.; funding acquisition, D.D.

Funding: This research was funded by Ghent University through the GOA-ENCLOSE project (BOF15/GOA/007).

Conflicts of Interest: The authors declare no conflict of interest.

Abbreviations

EPA	Energy per arriving adatom
P	Argon Pressure
d	Target-to-substrate distance
D	Domain Size
t	Layer thickness
FWHM	Full width at half maximum
RF	Radio-frequent
DC	Direct-current
RCA	Radio Corporation of America
MS	Mayadas-Shatzkes
PID	Proportional-integral-derivative
XRD	X-ray diffraction
TEM	Transmission electron microscope
AFM	Atomic force microscopy
EBSD	Electron backscatter diffraction

References

1. Laugier, M. The construction and use of thin film thermocouples for the measurement of surface temperature: Applications to substrate temperature determination and thermal bending of a cantilevered plate during film deposition. *Thin Solid Films* **1980**, *67*, 163–170. [CrossRef]
2. Kreider, K.G. Sputtered high temperature thin film thermocouples. *J. Vac. Sci. Technol. A Vac. Surf. Films* **1993**, *11*, 1401–1405. [CrossRef]
3. Scarioni, L.; Castro, E. Thermoelectric power in thin film Fe–CuNi alloy (type-J) couples. *J. Appl. Phys.* **2000**, *87*, 4337–4339. [CrossRef]
4. Chopra, K.; Bahl, S.; Randlett, M. Thermopower in thin-film copper—Constantan couples. *J. Appl. Phys.* **1968**, *39*, 1525–1528. [CrossRef]

5. Guo, H.; Jiang, J.Y.; Liu, J.X.; Nie, Z.H.; Ye, F.; Ma, C.F. Fabrication and Calibration of Cu-Ni Thin Film Thermocouples. *Adv. Mater. Res.* **2012**, *512–515*, 2068–2071. [CrossRef]
6. Sadat, S.; Tan, A.; Chua, Y.J.; Reddy, P. Nanoscale thermometry using point contact thermocouples. *Nano Lett.* **2010**, *10*, 2613–2617. [CrossRef] [PubMed]
7. Kim, K.; Jeong, W.; Lee, W.; Reddy, P. Ultra-high vacuum scanning thermal microscopy for nanometer resolution quantitative thermometry. *Acs Nano* **2012**, *6*, 4248–4257. [CrossRef]
8. Kim, K.; Song, B.; Fernández-Hurtado, V.; Lee, W.; Jeong, W.; Cui, L.; Thompson, D.; Feist, J.; Reid, M.H.; García-Vidal, F.J.; et al. Radiative heat transfer in the extreme near field. *Nature* **2015**, *528*, 387. [CrossRef]
9. Liu, H.; Sun, W.; Xiang, A.; Shi, T.; Chen, Q.; Xu, S. Towards on-chip time-resolved thermal mapping with micro-/nanosensor arrays. *Nanoscale Res. Lett.* **2012**, *7*, 484. [CrossRef]
10. Li, G.; Wang, Z.; Mao, X.; Zhang, Y.; Huo, X.; Liu, H.; Xu, S. Real-time two-dimensional mapping of relative local surface temperatures with a thin-film sensor array. *Sensors* **2016**, *16*, 977. [CrossRef]
11. Guk, E.; Ranaweera, M.; Venkatesan, V.; Kim, J.S. Performance and durability of thin film thermocouple array on a porous electrode. *Sensors* **2016**, *16*, 1329. [CrossRef] [PubMed]
12. Meredith, R.D.; Wrbanek, J.D.; Fralick, G.C.; Greer, L.C.; Hunter, G.W.; Chen, L. Design and operation of a fast, thin-film thermocouple probe on a turbine engine. In Proceedings of the 50th AIAA/ASME/SAE/ASEE Joint Propulsion Conference, Cleveland, OH, USA, 28–30 July 2014; p. 3923.
13. Zeng, Q.Y.; Hong, T.; Chen, L.; Cui, Y.X. Magnetron sputtering of NiCr/NiSi thin-film thermocouple sensor for temperature measurement when machining chemical explosive material. *Key Eng. Mater.* **2011**, *467*, 134–139. [CrossRef]
14. Geballe, T.; Hull, G. Seebeck effect in silicon. *Phys. Rev.* **1955**, *98*, 940. [CrossRef]
15. Gayner, C.; Kar, K.K. Recent advances in thermoelectric materials. *Prog. Mater. Sci.* **2016**, *83*, 330–382. [CrossRef]
16. Marshall, R.; Atlas, L.; Putner, T. The preparation and performance of thin film thermocouples. *J. Sci. Instrum.* **1966**, *43*, 144. [CrossRef]
17. Lin, S.F.; Leonard, W.F. Thermoelectric power of thin gold films. *J. Appl. Phys.* **1971**, *42*, 3634–3639. [CrossRef]
18. Yu, H.Y.; Leonard, W.F. Thermoelectric power of thin silver films. *J. Appl. Phys.* **1973**, *44*, 5324–5327. [CrossRef]
19. Leonard, W.F.; Yu, H. Thermoelectric power of thin copper films. *J. Appl. Phys.* **1973**, *44*, 5320–5323. [CrossRef]
20. Angadi, M.A.; Ashrit, P.V. Thermoelectric effect in ytterbium and samarium films. *J. Phys. D Appl. Phys.* **1981**, *14*, L125–L128. [CrossRef]
21. Angadi, M.; Udachan, L. Thermoelectric power measurements in thin tin films. *J. Phys. D Appl. Phys.* **1981**, *14*, L103. [CrossRef]
22. Angadi, M.A.; Shivaprasad, S.M. Thermoelectric power measurements in thin palladium films. *J. Mater. Sci. Lett.* **1982**, *1*, 65–66. [CrossRef]
23. Angadi, M.A.; Udachan, L.A. Thermopower measurements in chromium films. *J. Mater. Sci. Lett.* **1982**, *1*, 539–541. [CrossRef]
24. Cougnon, F.; Dulmaa, A.; Dedoncker, R.; Galbadrakh, R.; Depla, D. Impurity dominated thin film growth. *Appl. Phys. Lett.* **2018**, *112*, 221903. [CrossRef]
25. Guan, A.; Wang, H.; Jin, H.; Chu, W.; Guo, Y.; Lu, G. An experimental apparatus for simultaneously measuring Seebeck coefficient and electrical resistivity from 100 K to 600 K. *Rev. Sci. Instrum.* **2013**, *84*, 043903. [CrossRef]
26. Barber, Z.; Somekh, R. Magnetron sputtering of Cu55Ni45. *Vacuum* **1984**, *34*, 991–994. [CrossRef]
27. Gierczak, M.; Prażmowska-Czajka, J.; Dziedzic, A. Thermoelectric mixed thick-/thin film microgenerators based on constantan/silver. *Materials* **2018**, *11*, 115. [CrossRef]
28. Gall, D. Electron mean free path in elemental metals. *J. Appl. Phys.* **2016**, *119*, 085101. [CrossRef]
29. Sondheimer, E.H. The mean free path of electrons in metals. *Adv. Phys.* **1952**, *1*, 1–42. [CrossRef]
30. Mayer, H. Recent developments in conduction phenomena in thin metal films. In *Structure and Properties of Thin Films*; Neugebauer, C.A., Newkirk, J.W., Eds; Wiley: New York, NY, USA, 1959; p. 225.
31. Zhang, X.; Choi, H.; Datta, A.; Li, X. Design, fabrication and characterization of metal embedded thin film thermocouples with various film thicknesses and junction sizes. *J. Micromech. Microeng.* **2006**, *16*, 900. [CrossRef]

32. Rao, V.N.; Mohan, S.; Reddy, P.J. Electrical resistivity, TCR and thermoelectric power of annealed thin copper films. *J. Phys. D Appl. Phys.* **1976**, *9*, 89. [CrossRef]
33. Thakoor, A.; Suri, R.; Suri, S.; Chopra, K. Electron transport properties of copper films. II. Thermoelectric power. *J. Appl. Phys.* **1975**, *46*, 4777–4783. [CrossRef]
34. Rao, V.N.; Mohan, S.; Reddy, P.J. The size effect in the thermoelectric power of silver films. *Thin Solid Films* **1977**, *42*, 283–289. [CrossRef]
35. Hubin, M.; Gouault, J. Resistivity and thermoelectric power between −100 °C and +100 °C of gold and silver thin films formed and studied in ultrahigh vacuum. *Thin Solid Films* **1974**, *24*, 311–331. [CrossRef]
36. De, D.; Bandyopadhyay, S.K.; Chaudhuri, S.; Pal, A.K. Thermoelectric power of aluminum films. *J. Appl. Phys.* **1983**, *54*, 4022–4027. [CrossRef]
37. Boyer, A.; Cisse, E. Properties of thin film thermoelectric materials: Application to sensors using the Seebeck effect. *Mater. Sci. Eng. B* **1992**, *13*, 103–111. [CrossRef]
38. Das, V.D.; Soundararajan, N. Size and temperature effects on the Seebeck coefficient of thin bismuth films. *Phys. Rev. B* **1987**, *35*, 5990. [CrossRef]
39. Mikolajczak, P.; Piasek, W.; Subotowicz, M. Thermoelectric power in bismuth thin films. *Phys. Status Solidi A* **1974**, *25*, 619–628. [CrossRef]
40. Bourque-Viens, A.; Aimez, V.; Taberner, A.; Nielsen, P.; Charette, P.G. Modelling and experimental validation of thin-film effects in thermopile-based microscale calorimeters. *Sens. Actuators A Phys.* **2009**, *150*, 199–206. [CrossRef]
41. Mayadas, A.; Shatzkes, M. Electrical-resistivity model for polycrystalline films: The case of arbitrary reflection at external surfaces. *Phys. Rev. B* **1970**, *1*, 1382. [CrossRef]
42. Hofer, A.; Schlacher, J.; Keckes, J.; Winkler, J.; Mitterer, C. Sputtered molybdenum films: Structure and property evolution with film thickness. *Vacuum* **2014**, *99*, 149–152. [CrossRef]
43. Braeckman, B.; Misják, F.; Radnóczi, G.; Caplovicova, M.; Djemia, P.; Tetard, F.; Belliard, L.; Depla, D. The nanostructure and mechanical properties of nanocomposite Nb_x-CoCrCuFeNi thin films. *Scr. Mater.* **2017**, *139*, 155–158. [CrossRef]
44. Xin, Z.; Xiao-Hui, S.; Dian-Lin, Z. Thickness dependence of grain size and surface roughness for dc magnetron sputtered Au films. *Chin. Phys. B* **2010**, *19*, 086802. [CrossRef]
45. Srolovitz, D.J.; Battaile, C.C.; Li, X.; Butler, J.E. Simulation of faceted film growth in two-dimensions: Microstructure, morphology and texture. *Acta Mater.* **1999**, *47*, 2269–2281.
46. Song, X.; Liu, G. Computer simulation of normal grain growth in polycrystalline thin films. *J. Mater. Sci.* **1999**, *34*, 2433–2436. [CrossRef]
47. Thijssen, J. Simulations of polycrystalline growth in 2+1 dimensions. *Phys. Rev. B* **1995**, *51*, 1985. [CrossRef]
48. Van der Drift, A. Evolutionary selection, a principle governing growth orientation in vapour-deposited layers. *Philips Res. Rep.* **1967**, *22*, 267.
49. Barmak, K.; Darbal, A.; Ganesh, K.J.; Ferreira, P.J.; Rickman, J.M.; Sun, T.; Yao, B.; Warren, A.P.; Coffey, K.R. Surface and grain boundary scattering in nanometric Cu thin films: A quantitative analysis including twin boundaries. *J. Vac. Sci. Technol. A Vac. Surf. Films* **2014**, *32*, 061503. [CrossRef]
50. Sun, T.; Yao, B.; Warren, A.P.; Barmak, K.; Toney, M.F.; Peale, R.E.; Coffey, K.R. Dominant role of grain boundary scattering in the resistivity of nanometric Cu films. *Phys. Rev. B* **2009**, *79*, 041402. [CrossRef]
51. Zhang, X.; Misra, A. Residual stresses in sputter-deposited copper/330 stainless steel multilayers. *J. Appl. Phys.* **2004**, *96*, 7173–7178. [CrossRef]
52. Yu, S.; Li, L.; Lyu, X.; Zhang, W. Preparation and investigation of nano-thick FTO/Ag/FTO multilayer transparent electrodes with high figure of merit. *Sci. Rep.* **2016**, *6*, 20399. [CrossRef]
53. Ding, G.; Clavero, C. Silver-based low-emissivity coating technology for energy-saving window applications. In *Modern Technologies for Creating the Thin-film Systems and Coatings*; Nikitenkov, N., Ed.; InTechOpen: London, UK, 2017; pp. 409–431.
54. Shamsutdinov, N.; Sloof, W.; Böttger, A. A method for the experimental determination of surface photoemission core-level shifts for 3d transition metals. *J. Appl. Phys.* **2005**, *98*, 014908. [CrossRef]
55. Krist, T.; Teichert, A.; Meltchakov, E.; Vidal, V.; Zoethout, E.; Müllender, S.; Bijkerk, F. Stress reduction in multilayers used for X-Ray and neutron optics. In *Modern Developments in X-Ray and Neutron Optics*; Erko, A., Idir, M., Eds.; Springer: Heidelberg, Germany, 2008; pp. 371–388.

56. Cougnon, F.; Schramm, I.; Depla, D. On the electrical properties of sputter deposited thin films: The role of energy and impurity flux. *Thin Solid Films* **2019**, submitted for publication.
57. Wanarattikan, P.; Jitthammapirom, P.; Sakdanuphab, R.; Sakulkalavek, A. Effect of grain size and film thickness on the thermoelectric properties of flexible Sb_2Te_3 thin films. *Adv. Mater. Sci. Eng.* **2019**, *2019*, 6954918. [CrossRef]
58. Mayadas, A.F.; Janak, J.F.; Gangulee, A. Resistivity of Permalloy thin films. *J. Appl. Phys.* **1974**, *45*, 2780. [CrossRef]
59. Sun, T.; Yao, B.; Warren, A.P.; Barmak, K.; Toney, M.F.; Peale, R.E.; Coffey, K.R. Surface and grain-boundary scattering in nanometric Cu films. *Phys. Rev. B* **2010**, *81*, 155454. [CrossRef]
60. Zhang, W.; Brongersma, S.; Heylen, N.; Beyer, G.; Vandervorst, W.; Maex, K. Geometry effect on impurity incorporation and grain growth in narrow copper lines. *J. Electrochem. Soc.* **2005**, *152*, C832–C837. [CrossRef]
61. Birkett, M.; Penlington, R. Electrical resistivity of CuAlMo thin films grown at room temperature by dc magnetron sputtering. *Mater. Res. Express* **2016**, *3*, 075021. [CrossRef]
62. Deschacht, D.; Boyer, A.; Groubert, E. The thermoelectric power of polycrystalline semimetal films. *Phys. Status Solidi A* **1982**, *71*, K205–K209. [CrossRef]
63. Beensh-Marchwicka, G.; Osadnik, S.; Prociów, E.; Mielcarek, W. Structure and morphology of Ge(Au) sputtered films with useful Seebeck effect. *Vacuum* **1998**, *50*, 207–210. [CrossRef]
64. Arab Pour Yazdi, M.; Martin, N.; Petitot, C.; Neffaa, K.; Palmino, F.; Cherioux, F.; Billard, A. Influence of sputtering parameters on structural, electrical and thermoelectric properties of Mg–Si coatings. *Coatings* **2018**, *8*, 380. [CrossRef]
65. Gordillo, G.; Mesa, F.; Calderón, C. Electrical and morphological properties of low resistivity Mo thin films prepared by magnetron sputtering. *Braz. J. Phys.* **2006**, *36*, 982–985. [CrossRef]
66. Chan, K.; Teo, B. Effect of Ar pressure on grain size of magnetron sputter-deposited Cu thin films. *IET Sci. Meas. Technol.* **2007**, *1*, 87–90. [CrossRef]
67. Qiu, H.; Wang, F.; Wu, P.; Pan, L.; Tian, Y. Structural and electrical properties of Cu films deposited on glass by DC magnetron sputtering. *Vacuum* **2002**, *66*, 447–452. [CrossRef]
68. Van Aeken, K.; Mahieu, S.; Depla, D. The metal flux from a rotating cylindrical magnetron: A Monte Carlo simulation. *J. Phys. D Appl. Phys.* **2008**, *41*, 205307. [CrossRef]
69. Hoffman, D.; Thornton, J.A. Internal stresses in Cr, Mo, Ta, and Pt films deposited by sputtering from a planar magnetron source. *J. Vac. Sci. Technol.* **1982**, *20*, 355–358. [CrossRef]
70. Depla, D.; Mahieu, S. *Reactive Sputter Deposition*; Springer: Heidelberg, Germany, 2008.
71. Liu, H.; Sun, W.; Xu, S. An extremely simple thermocouple made of a single layer of metal. *Adv. Mater.* **2012**, *24*, 3275–3279. [CrossRef]

© 2019 by the authors. Licensee MDPI, Basel, Switzerland. This article is an open access article distributed under the terms and conditions of the Creative Commons Attribution (CC BY) license (http://creativecommons.org/licenses/by/4.0/).

Article

Gas Sensing with Nanoplasmonic Thin Films Composed of Nanoparticles (Au, Ag) Dispersed in a CuO Matrix

Manuela Proença *, Marco S. Rodrigues, Joel Borges * and Filipe Vaz

Centro de Física da Universidade do Minho, Campus de Gualtar, 4710-057 Braga, Portugal; marcopsr@gmail.com (M.S.R.); fvaz@fisica.uminho.pt (F.V.)
* Correspondence: manuelaproenca12@gmail.com (M.P.); joelborges@fisica.uminho.pt (J.B.); Tel.: +351-253-510-471 (J.B.)

Received: 6 May 2019; Accepted: 23 May 2019; Published: 25 May 2019

Abstract: Magnetron sputtered nanocomposite thin films composed of monometallic Au and Ag, and bimetallic Au-Ag nanoparticles, dispersed in a CuO matrix, were prepared, characterized, and tested, which aimed to find suitable nano-plasmonic platforms capable of detecting the presence of gas molecules. The Localized Surface Plasmon Resonance phenomenon, LSPR, induced by the morphological changes of the nanoparticles (size, shape, and distribution), and promoted by the thermal annealing of the films, was used to tailor the sensitivity to the gas molecules. Results showed that the monometallic films, Au:CuO and Ag:CuO, present LSPR bands at ~719 and ~393 nm, respectively, while the bimetallic Au-Ag:CuO film has two LSPR bands, which suggests the presence of two noble metal phases. Through transmittance-LSPR measurements, the bimetallic films revealed to have the highest sensitivity to the refractive index changes, as well as high signal-to-noise ratios, respond consistently to the presence of a test gas.

Keywords: thin films; magnetron sputtering; microstructure; noble metal nanoparticles; CuO matrix; localized surface plasmon resonance; gas sensor

---

1. Introduction

Nanocomposite thin films, containing noble metal nanoparticles embedded in an oxide matrix, have been a subject of considerable interest for optical gas sensing due to their localized surface plasmon resonance (LSPR) properties [1,2]. Surface plasmons are coherent oscillations of free electrons excited by an electromagnetic field at the boundaries between a metal and a dielectric. They can propagate along the surface of the conductor, which are designated by surface plasmon polaritons, or be confined to metallic nanoparticles or nanostructures, in which case, are denominated as localized surface plasmons [3–5]. LSPR can give rise to strong absorption bands, the enhancement of the electromagnetic field near the nanoparticles, and the appearance of scattering to the far field [6–10]. Since its discovery, there have been significant advances in both theoretical and experimental investigations of surface plasmons, which led to the development of new modelling methods that contribute to the understanding of the morphology and to the calculation of the optical properties of nano-plasmonic systems [8,11,12].

The two most well studied plasmonic metals are gold (Au) and silver (Ag). They exhibit LSPR bands within the visible spectrum due to the energy levels of d-d transitions, being used in various applications involving color [13,14] as well as in sensing due to their relatively high refractive index sensitivity [15,16]. Since Ag nanoparticles present the sharpest and strongest bands among all metals, they are associated to higher sensitivity factors than Au. However, Au nanoparticles are more frequently selected for sensing applications due to their lower toxicity, inert nature (less prone to oxidation), and stability [17,18]. On the other hand, Ag-Au bimetallic nanoparticles have attracted particular attention due to their corresponding monometallic counterparts, which may allow further improvements on their set of properties [19–21], especially the optical behavior [16,21,22]. In fact, they are relatively easy to prepare since both metals have a face-centred cubic structure and similar lattice constants. However, it is known that the synthesis method can result in alloyed bimetallic nanoparticles [23,24], core-shell [25], and even Janus systems [26]. These features are determined by the Au/Ag ratio in the bimetallic nanostructure, which leads to different optical properties. From the alloy formation of Au-Ag bimetallic nanoparticles, only one LSPR band results between the peaks of the constituting monometallic nanoparticles, while a mixed system originates two plasmonic bands, as reported in different works [3,22,23,27].

Diverse noble metal compositions dispersed in a dielectric matrix and different microstructures and nanostructures might be developed, which originates different LSPR bands, since their curvature and position are strongly dependent on different factors such as the composition, size, shape, and distribution of the nanoparticles, which are also sensitive to changes of the refractive index of the surrounding dielectric medium where they are dispersed [1,3]. Hence, the basis of the plasmonic bio/chemical sensors is established by the dependence of the LSPR band on the surrounding refractive index [2,16,28]. One of the advantages of using LSPR phenomenon for optical gas sensors in contrast to Surface Plasmon Resonance (SPR) systems is the fact that the first ones have a much higher potential to be sensitive to the extremely low refractive index changes such as those induced by gas molecules [2,29,30], since the plasmon decay length in LSPR is much lower than in SPR [31]. Furthermore, LSPR-based sensors are basically supported by nanoparticles that can be directly coupled to light, while the SPR-based sensors are dependent on prisms, optical fibers, or gratings to be coupled with light [30,32,33].

For the LSPR gas detection by refractive index changes to be functional, the production of highly sensitive plasmonic thin films is required, but the development of a high-resolution spectroscopy system to measure extremely small LSPR peak shifts is a fact that has been hampering the research on this area [2,34]. Hence, in order to optimize the sensitivity of the films, previous studies of the LSPR sensing response have been made by using two liquids with a relatively large refractive index difference [28,35,36], which allowed us to estimate the refractive index sensitivity (RIS) [37–39].

The present work proposes a reliable and effective possibility of sensitive thin films, suitable to be used as optical sensors. Such (nanocomposite) thin films are based on Au and/or Ag nanoparticles, dispersed in a semiconductor copper oxide (CuO) matrix, Au:CuO, Ag:CuO, and Au-Ag:CuO, deposited by reactive DC magnetron sputtering. The use of a pure copper target containing gold and/or silver pellets on its surface, avoids the use of a second cathode [40,41], with evident economic advantages [3,28]. After the preparation of the thin films, a thermal annealing treatment was performed in order to promote the necessary nanostructural changes in the noble metal nanoparticles, and dielectric matrix, which enabled the manifestation of the LSPR behavior, and, consequently, turned the thin films sensitive to the gas molecules. The composition and morphology of the thin films were studied and correlated with the LSPR responses. LSPR sensing tests were performed through transmittance measurements in a custom-made optical vacuum system, which incorporates a gas flow cell. The sensitivity of the different films to the presence of O_2 gas was also calculated and compared between them.

2. Materials and Methods

Thin films of Au:CuO, Ag:CuO, and Au-Ag:CuO manifesting LSPR behavior were produced by a two-step process, involving deposition of the thin films and posterior thermal treatment. For the depositions, two different types of substrates were used including Si (Boron doped, p type, <100> orientation, 525 µm thick) for chemical and (micro)structural characterization purposes and SiO_2 (fused silica) for optical spectra measurements. Before the depositions and in order to clean and activate the surface of the substrates, plasma treatments were performed by a Low-Pressure Plasma Cleaner by Diener Electronic equipped with a 40 kHz RF generator (Zepto Model, Ebhausen, Germany) [42], applying a power of 100 W. The substrates were first cleaned with O_2 plasma (80 Pa, for 5 min), and then activated with Ar plasma (80 Pa, for 15 min).

The films were deposited by reactive (DC) magnetron sputtering during 60 s in order to produce films with thicknesses around ~50 nm. As illustrated in Figure 1a, the above-mentioned substrates were then placed in a grounded hexagonal holder, rotating at 16 rpm and 7 cm far from the cathode. The latter is a rectangular copper target (200 × 100 × 6 mm^3, 99.99% purity), where gold and/or silver pellets (surface area of 960 mm^2 and 0.5 mm thick) were placed symmetrically on its preferential erosion zone. The base pressure was below 5×10^{-4} Pa, while the target potential was limited to 500 V, and the applied current was 3.25 mA/cm^2. The discharge was ignited in a gas atmosphere composed of Ar (3.5×10^{-1} Pa) and O_2 (2×10^{-2} Pa). Then, in order to promote the nanoparticles' growth, the films were subjected to thermal treatments in-air, up to a maximum temperature of 700 °C, according to what was previously studied and published by the group [1,28]. The heating ramp used was 5 °C/min and the isothermal period was 5 h, which cooled down freely inside the furnace, before reaching room temperature.

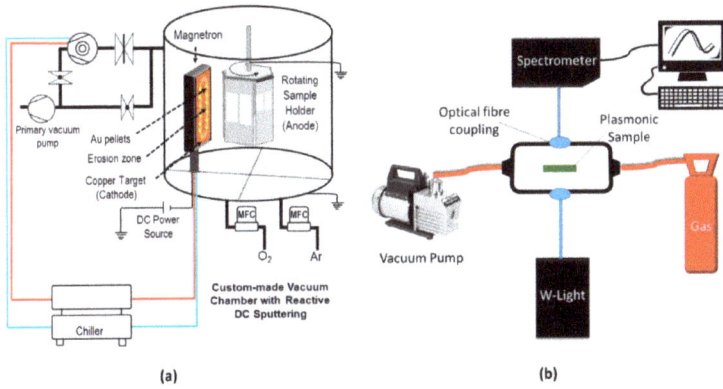

Figure 1. Simplified representation of the reactive DC magnetron sputtering system (**a**) and the custom-made system for transmittance-LSPR (T-LSPR) measurement in a controlled atmosphere (**b**).

The atomic composition of the films was studied by Rutherford Backscattering Spectrometry (RBS) using a Van de Graaff accelerator, a standard detector, placed at 140°, and two pin-diode detectors located symmetrically to each other, both placed at a 165° scattering angle respective to the beam direction. Spectra were collected using 2.0 MeV $4He^+$, and 1.45 MeV $1H^+$ beams at normal incidence and the data was analyzed with the IBA DataFurnace NDF v9.6i code [43].

The morphology of the films' surface was studied by a Dual Beam Scanning Electron Microscope, SEM/FIB FEI Helios 600i (Hillsboro, OR, USA), using a backscatter electron detector. The surface micrographs were analyzed using MATLAB software (version R2018a), by calculating the Feret diameter, the aspect ratio, and the nearest neighbor of the contrasted nanoparticles. The MATLAB algorithm included the locally adaptive threshold function "adaptthresh." After the binarization and scaling of the SEM images, the nanoparticles were analysed using the "regionprops" and "bwboundaries" functions.

The films' gas sensitivity was investigated by monitoring the LSPR band in the presence of O_2 (atmospheric pressure), in comparison to a low vacuum pressure. Real-time measurements were performed in a custom-made system (Figure 1b), composed of two main parts: the optical components and a vacuum system. The optical system allows the measurement of the optical (transmittance) spectrum of the sample, using a tungsten lamp and a modular spectrometer by Ocean Optics (HR4000, Edinburgh, UK). Optical fibers were used to connect those components to the flow cell, where the sample is placed. A vacuum pump was used to produce a "primary" vacuum (~40 Pa) inside the flow cell and then O_2 was introduced at atmospheric pressure for 120 s. Several vacuum/O_2 cycles were

3. Results

3.1. Thin Films Characterization

The atomic concentration profiles of the thin films were determined by RBS (Figure 2). The as-deposited CuO matrix (solid lines), and the CuO matrix with thermal treatment at 700 °C (dash lines), are represented in Figure 2a, while the as-deposited nanocomposite films are displayed in Figure 2b–d. According to the RBS analysis, all the as-deposited thin films were found to have a roughly constant atomic concentration across their thickness, even after the annealing process for the case of the pure matrix. Moreover, elemental concentration results revealed that the matrix of the as-deposited films is not fully CuO stoichiometric, since the atomic ratio C_O/C_{Cu} is always different from but close to 1. However, as soon as the film is subjected to thermal annealing, it seems that the CuO matrix becomes stoichiometric, which can be observed by the corresponding RBS profile (Figure 2a), where Cu and O concentrations were estimated to be about 50.0 ± 0.5 at.% and 50 ± 3 at.%, respectively. Thus, when the films are subjected to thermal treatment in air, the chemical composition may change in relation to the as-deposited films due to oxygen incorporation [44,45], as previously verified [1,28]. The atomic concentration of noble metals into the CuO matrix was determined to be about C_{Au} = 15.0 ± 0.5 at.% (Au:CuO), C_{Ag} = 17.7 ± 0.5 at.% (Ag:CuO), and C_{Au} = 6.7 ± 0.5 at.%, C_{Ag} = 8.0 ± 0.5 at.% (Au-Ag:CuO). These were the compositions of the thin films used for LSPR sensitivity tests.

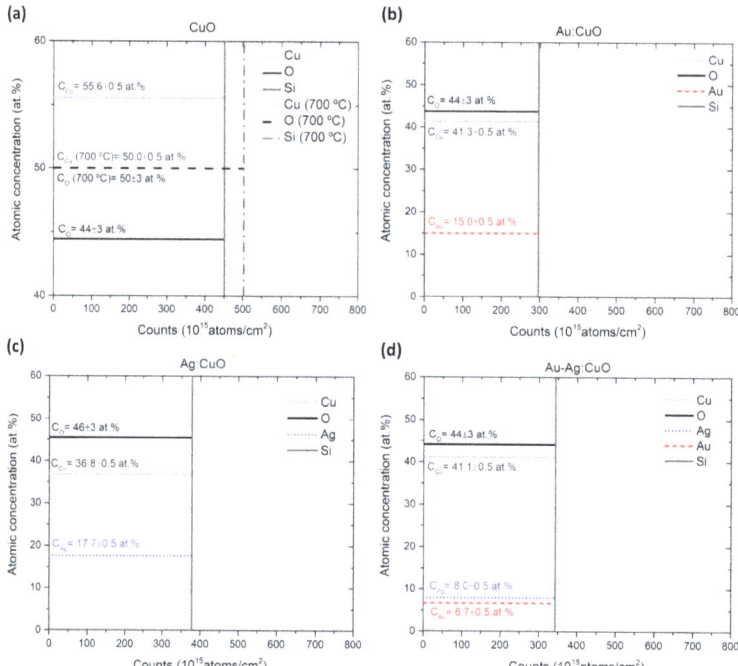

Figure 2. Atomic concentration (at.%) of the different elements present in the as-deposited CuO matrix (solid lines (**a**)), in the CuO matrix with annealing at 700 °C (dash lines (**a**)), and in the as-deposited samples of Au:CuO (**b**), Ag:CuO (**c**), and Au-Ag:CuO (**d**) films deposited with a pellets' area of 960 mm², obtained by the RBS data analyzed with the code IBA DataFurnace NDF v9.6i [43].

The CuO matrix annealed at 700 °C presents a polycrystalline structure with well-defined grain boundaries, as observed in the SEM micrograph displayed in Figure 3a. In addition, the optical transmittance spectrum (Figure 3b) reveals a semi-transparent CuO matrix in the visible range, with a progressive increase of transmittance for higher wavelengths, which is a feature that is in agreement with the literature [46].

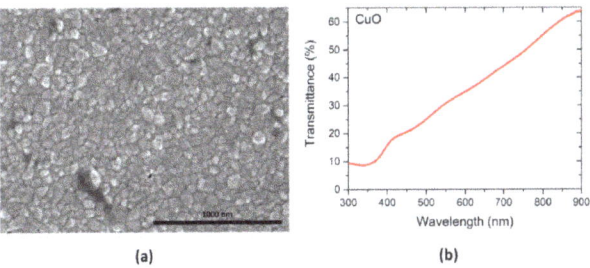

Figure 3. Top-view SEM micrograph of the CuO matrix annealed at 700 °C (**a**) and the respective optical transmittance spectrum (**b**).

The microstructural analysis of the annealed plasmonic thin films revealed the presence of noble nanoparticles (bright spots) in the different nanocomposite thin films ((a) and (b) in Figures 4–6), which suggests that the growth of nanoparticles might be facilitated by easier diffusion of Au and Ag atoms through grain boundaries of the CuO matrix. The Au:CuO (Figure 4) film is the one that presents the highest nanoparticles' density at the surface (127 μm^{-2}) with an average size of about 33 nm (Figure 4c). Moreover, the Au nanoparticles are relatively close to each other (Figure 4d) and they are presumably spherical since their aspect ratio distribution is narrow and close to 1, as seen in Figure 4e.

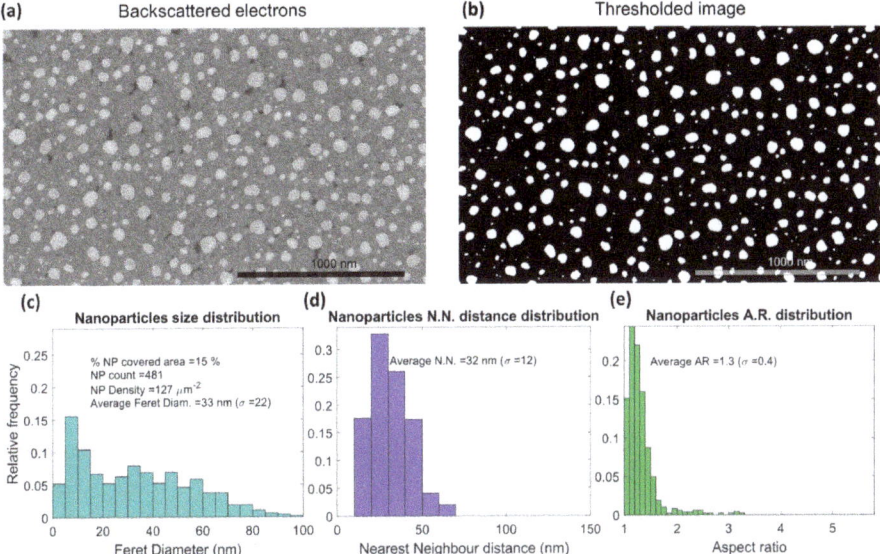

Figure 4. Au nanoparticle distribution analysis, using a MATLAB algorithm: (**a**) top-view SEM micrograph with atomic weight contrast and (**b**) processed and threshold micrograph showing the exposed Au nanoparticles, with 15% Au coverage area. Distribution histograms are displayed in (**c**) for the nanoparticles Feret diameter, (**d**) for the nearest neighbor distance, and (**e**) for the aspect ratio.

Regarding the Ag:CuO film (Figure 5), the average size of Ag nanoparticles was estimated to be 15 nm. However, the nanoparticles' density at the surface (69 µm^2) is much smaller than in the other films (Figure 5c), which leads to the highest distance between the nanoparticles (Figure 5d). In fact, the formation of islands of Ag (micro-sized agglomerates with parallelepiped shape) was observed on the surface of the film (not shown here) [28]. This explains the low amount of nanosized Ag particles, which is a behavior that was not expected when taking into account the relatively high Ag atomic concentration determined for the as-deposited film.

The Au-Ag:CuO film (Figure 6) presents values between those belonging to the monometallic counterparts (Figures 4 and 5). It presents a density of Au-Ag nanoparticles at the surface of 100 µm^{-2}, with an average size estimated to be 30 nm (Figure 6c). Moreover, the nearest neighbor distance distribution is broader than in the Au film and narrower than in the Ag film. Moreover, this system shows the widest aspect ratio distribution, with an average value of 1.5, which proves that both spherical and irregular nanoparticles are present in the film's surface.

The different microstructures achieved by the films with the thermal treatment originated different optical transmittance responses, as shown in Figure 7. The high Au nanoparticles' density at the surface and their quasi-spherical shape, observed in the Au:CuO film (Figure 4), gave rise to a well-defined and sharp transmittance LSPR (T-LSPR) band at ~719 nm (Figure 7a), with a high transmittance amplitude, at about 15 percentage points (i.e., the difference between the maximum and the minimum band's peak).

On the other hand, a T-LSPR band was also observed for the Ag:CuO film (Figure 7b), appearing at shorter wavelengths (~393 nm) as is typical of the Ag nanoparticles [3,28]. However, despite the narrow shape, due to its slightly larger nanoparticle aspect ratio distribution, the LSPR band is also less intense since the number of Ag nanoparticles at the surface is scarce, which presents only a transmittance amplitude of ~10 percentage points.

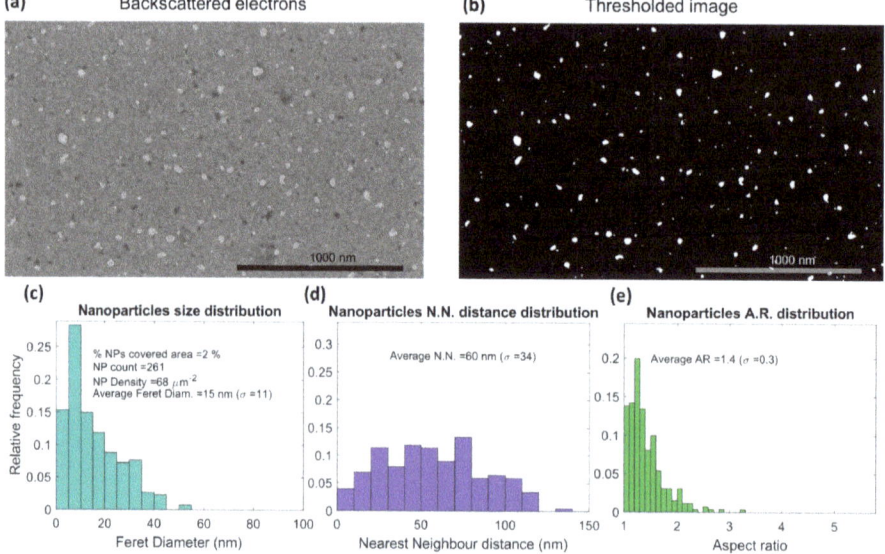

Figure 5. Ag nanoparticle distribution analysis, using a MATLAB algorithm: (**a**) top-view SEM micrograph with atomic weight contrast and (**b**) processed and threshold micrograph showing the exposed Ag nanoparticles, with 2% Ag coverage area. Distribution histograms are displayed in (**c**) for nanoparticles Feret diameter, (**d**) for the nearest neighbor distance, and (**e**) for the aspect ratio.

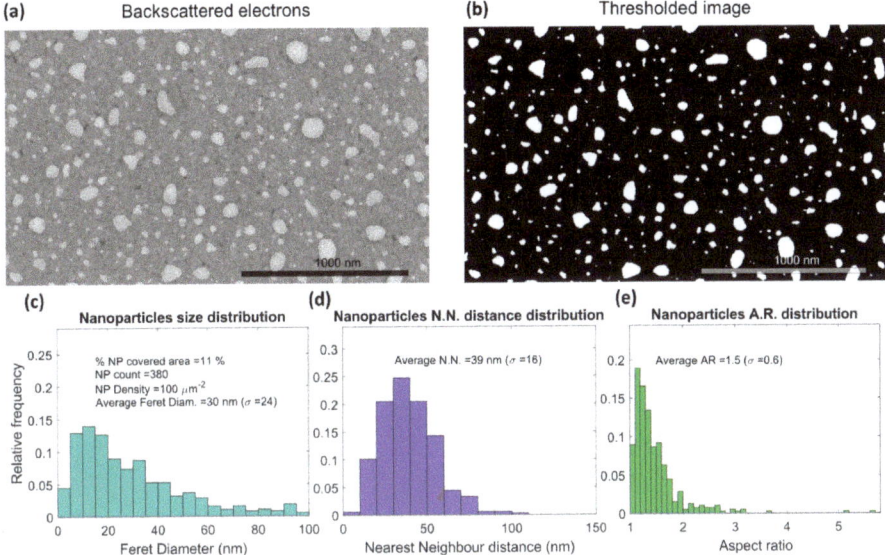

Figure 6. Au-Ag nanoparticle distribution analysis, using a MATLAB algorithm: (**a**) top-view SEM micrograph with atomic weight contrast and (**b**) processed and threshold micrograph showing the exposed Au-Ag nanoparticles, with 11% Au-Ag coverage area. Distribution histograms are displayed in (**c**) for nanoparticles Feret diameter, (**d**) for the nearest neighbor distance, and (**e**) for the aspect ratio.

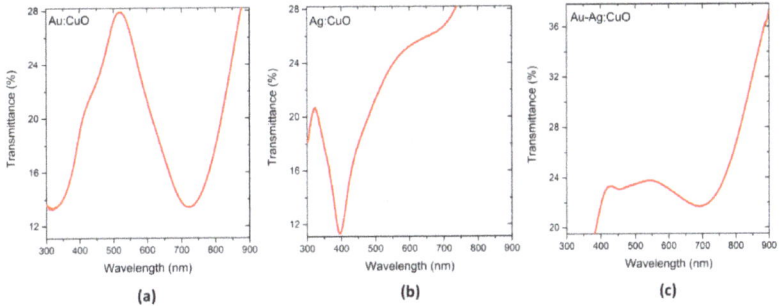

Figure 7. Transmittance spectra of the Au:CuO (**a**), Ag:CuO (**b**), and Au-Ag:CuO (**c**) thin films after in-air annealing.

Concerning the Au-Ag:CuO film, two shifted LSPR peaks are observed (~450 and 676 nm), even though the second one is much more pronounced (Figure 7c). The presence of two peaks might suggest the presence of separate phases of Ag and Au nanoparticles in these films, but since they are shifted from their initial positions, the formation of an alloy of Au-Ag bimetallic nanoparticles cannot be disregarded. Furthermore, as observed in Figure 6, the film presents both spherical and elongated nanoparticles, which contribute to the LSPR band widening and, therefore, appears much less intense.

3.2. Sensitivity Tests Using Exposure to O_2

In order to test the films' sensitivity to refractive index changes promoted by the presence and/or adsorption of gas molecules, they were exposed to a test gas (O_2). Figure 8 presents the LSPR peak position (transmittance) of the three systems, during five cycles under vacuum, and O_2 at atmospheric pressure. As expected from this type of sensor, the transmittance shift due to a change in the refractive index is typically very short, in the order of tenths of percentage points [2,47]. Anyway, it is possible to observe that the films responded consistently to the presence of the gas. The T-LSPR peak shifted to lower transmittances when the O_2 was introduced, which decreases by 0.35, 0.11, and 0.43 percentage points for the Au, Ag, and Au-Ag:CuO films, respectively. These results are consistent with what has been already published for Au-TiO$_2$ films, but with slightly higher sensitivities [2]. The Ag:CuO sample presents the lowest shift and, subsequently, the lowest signal-to-noise ratio (~3). This is believed to result from the morphology achieved after the annealing process (Figure 5), where the presence of Ag nanoparticles at the film's surface is scarce, which might hinder the film's sensitivity. Moreover, the presence of Au in the Au:CuO film and both Ag and Au nanoparticles in Au-Ag:CuO film seems to improve the film's response since a higher transmittance shift is observed when the test gas is introduced. In addition to show the highest transmittance shift, the Au-Ag:CuO film also presents the best signal-to-noise ratio (~123) even though the Au:CuO film has also a reasonable value of ~59. Furthermore, the peak shifts are reproducible every cycle when the test gas is introduced, which suggests that the eventual gas adsorption is reversible.

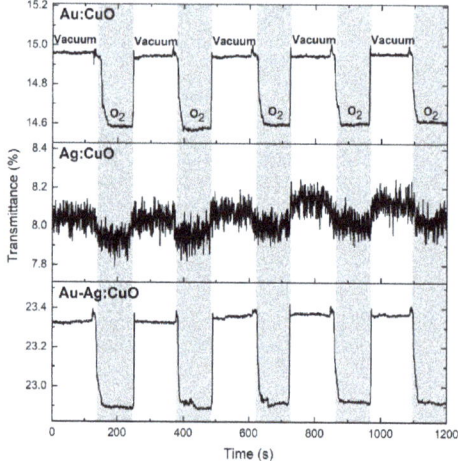

Figure 8. Variation of the LSPR peak position (transmittance minimum) of the Au:CuO, Ag:CuO, and Au-Ag:CuO films over time for five cycles of vacuum and O_2 atmosphere.

4. Conclusions

Au:CuO, Ag:CuO, and Au-Ag:CuO thin films with nanoplasmonic properties were produced in this work. The films were first deposited by magnetron sputtering for 1 min, using a Cu target with small metallic pellets (960 mm^2 pellet area) and a target potential limited to 500 V. Then, the different films were annealed up to 700 °C in order to promote the nanoparticles' growth and structural changes.

The composition analysis revealed the presence of reasonable amounts of noble metals in a CuO matrix, which becomes stoichiometric after a thermal treatment in air. Furthermore, the annealing induced structural and morphological changes that influenced the LSPR responses of the thin films. Due to the presence of spherical Au nanoparticles with high density at the surface, the Au:CuO film presented the most well-defined and pronounced transmittance LSPR band at ~719 nm, while the Ag:CuO film showed a narrower but less intense band at shorter wavelengths (~393 nm) due to the scarce number of Ag nanoparticles at the surface. However, despite the fact that the Au-Ag:CuO film has two T-LSPR peaks (~450 and 676 nm) with relatively low intensity, it showed to be the most sensitive to the refractive index changes, such as to the O_2 gas presence, followed after by Au:CuO and Ag:CuO films.

In conclusion, this work proves that the sensitivity of Au-Ag:CuO thin films to the test gas (O_2) can be improved by preparing bimetallic noble nanoparticles embedded in the CuO matrix. Hence, this configuration might be preferable to use for LSPR gas sensing.

Author Contributions: Conceptualization, J.B., F.V.; methodology, M.S.R., J.B.; software, M.S.R.; validation, M.P., M.S.R., J.B., F.V.; formal analysis, M.P., M.S.R., J.B., F.V.; investigation, M.P., M.S.R., J.B.; resources, J.B., F.V.; data curation, J.B., M.S.R., F.V.; writing—original Draft preparation, M.P., J.B.; writing—review and Editing, J.B., F.V.; visualization, M.P., M.S.R., J.B., F.V.; supervision, J.B., F.V.; project administration, F.V.; funding acquisition, J.B., F.V.

Funding: This research was funded by the Portuguese Foundation for Science and Technology (FCT) in the framework of the Strategic Funding UID/FIS/04650/2019; and by the projects NANOSENSING POCI-01-0145-FEDER-016902, with FCT reference PTDC/FIS-NAN/1154/2014; and project NANO4BIO POCI-01-0145-FEDER-032299, with FCT reference PTDC/FIS-MAC/32299/2017 supported this work. Manuela Proença acknowledges her PhD Scholarship from FCT, with reference SFRH/BD/137076/2018. Joel Borges acknowledges FCT for his Researcher Contract from project NANO4BIO, CTTI-149/18-CF(1). Marco S. Rodrigues acknowledges FCT for his PhD Scholarship, SFRH/BD/118684/2016.

Conflicts of Interest: The authors declare no conflict of interest.

References

1. Proença, M.; Borges, J.; Rodrigues, M.S.; Domingues, R.P.; Dias, J.P.; Trigueiro, J.; Bundaleski, N.; Teodoro, O.M.N.D.; Vaz, F. Development of Au/CuO nanoplasmonic thin films for sensing applications. *Surf. Coat. Technol.* **2018**, *343*, 178–185. [CrossRef]
2. Rodrigues, M.S.; Borges, J.; Proença, M.; Pedrosa, P.; Martin, N.; Romanyuk, K.; Kholkin, A.L.; Vaz, F. Nanoplasmonic response of porous Au-TiO_2 thin films prepared by oblique angle deposition. *Nanotechnology* **2019**, *30*, 22. [CrossRef]
3. Borges, J.; Ferreira, C.G.; Fernandes, J.P.C.; Rodrigues, M.S.; Proença, M.; Apreutesei, M.; Alves, E.; Barradas, N.P.; Moura, C.; Vaz, F. Thin films of Ag-Au nanoparticles dispersed in TiO_2: Influence of composition and microstructure on the LSPR and SERS responses. *J. Phys. D. Appl. Phys.* **2018**, *51*, 205102. [CrossRef]
4. Borges, J.; Kubart, T.; Kumar, S.; Leifer, K.; Rodrigues, M.S.; Duarte, N.; Martins, B.; Dias, J.P.; Cavaleiro, A.; Vaz, F. Microstructural evolution of Au/TiO_2 nanocomposite films: The influence of Au concentration and thermal annealing. *Thin Solid Films* **2015**, *580*, 77–88. [CrossRef]
5. Borges, J.; Pereira, R.M.S.; Rodrigues, M.S.; Kubart, T.; Kumar, S.; Leifer, K.; Cavaleiro, A.; Polcar, T.; Vasilevskiy, M.I.; Vaz, F. Broadband optical absorption caused by the plasmonic response of coalesced Au nanoparticles embedded in a TiO_2 matrix. *J. Phys. Chem. C* **2016**, *120*, 16931–16945. [CrossRef]
6. Ghosh, S.K.; Pal, T. Interparticle coupling effect on the surface plasmon resonance of gold nanoparticles: From theory to applications. *Chem. Rev.* **2007**, *107*, 4797–4862. [CrossRef]
7. Hutter, E.; Fendler, J.H. Exploitation of localized surface plasmon resonance. *Adv. Mater.* **2004**, *16*, 1685–1706. [CrossRef]
8. Toudert, J.; Simonot, L.; Camelio, S.; Babonneau, D. Advanced optical effective medium modeling for a single layer of polydisperse ellipsoidal nanoparticles embedded in a homogeneous dielectric medium: Surface plasmon resonances. *Phys. Rev. B* **2012**, *86*, 045415. [CrossRef]
9. Politano, A.; Formoso, V.; Chiarello, G. Dispersion and damping of gold surface plasmon. *Plasmonics* **2008**, *3*, 165–170. [CrossRef]
10. Pitarke, J.M.; Silkin, V.M.; Chulkov, E.V.; Echenique, P.M. Theory of surface plasmons and surface-plasmon polaritons. *Rep. Prog. Phys.* **2007**, *70*, 1. [CrossRef]
11. Scholl, J.A.; Koh, A.L.; Dionne, J.A. Quantum plasmon resonances of individual metallic nanoparticles. *Nature* **2012**, *483*, 421–427. [CrossRef]
12. Goyenola, C.; Gueorguiev, G.K.; Stafström, S.; Hultman, L. Fullerene-like CS_x: A first-principles study of synthetic growth. *Chem. Phys. Lett.* **2011**, *506*, 86–91. [CrossRef]
13. Rodrigues, M.S.; Borges, J.; Gabor, C.; Munteanu, D.; Apreutesei, M.; Steyer, P.; Lopes, C.; Pedrosa, P.; Alves, E.; Barradas, N.P.; et al. Functional behaviour of TiO_2 films doped with noble metals. *Surf. Eng.* **2016**, *32*, 554–561. [CrossRef]

14. Torrell, M.; Cunha, L.; Cavaleiro, A.; Alves, E.; Barradas, N.P.; Vaz, F. Functional and optical properties of Au:TiO$_2$ nanocomposite films: The influence of thermal annealing. *Appl. Surf. Sci.* **2010**, *256*, 6536–6542. [CrossRef]
15. Zhao, Y.; Yang, Y.; Cui, L.; Zheng, F.; Song, Q. Electroactive Au@Ag nanoparticles driven electrochemical sensor for endogenous H$_2$S detection. *Biosens. Bioelectron.* **2018**, *117*, 53–59. [CrossRef] [PubMed]
16. Ghodselahi, T.; Arsalani, S.; Neishaboorynejad, T. Synthesis and biosensor application of Ag@Au bimetallic nanoparticles based on localized surface plasmon resonance. *Appl. Surf. Sci.* **2014**, *301*, 230–234. [CrossRef]
17. Borges, J.; Buljan, M.; Sancho-Parramon, J.; Bogdanovic-Radovic, I.; Siketic, Z.; Scherer, T.; Kübel, C.; Bernstorff, S.; Cavaleiro, A.; Vaz, F.; et al. Evolution of the surface plasmon resonance of Au:TiO$_2$ nanocomposite thin films with annealing temperature. *J. Nanopart. Res.* **2014**, *16*, 2790. [CrossRef]
18. Petryayeva, E.; Krull, U.J. Localized surface plasmon resonance: Nanostructures, bioassays and biosensing—A review. *Anal. Chim. Acta* **2011**, *706*, 8–24. [CrossRef] [PubMed]
19. Cesca, T.; Michieli, N.; Kalinic, B.; Balasa, I.G.; Rangel-Rojo, R.; Reyes-Esqueda, J.A.; Mattei, G. Bidimensional ordered plasmonic nanoarrays for nonlinear optics, nanophotonics and biosensing applications. *Mater. Sci. Semicond. Process.* **2019**, *92*, 2–9. [CrossRef]
20. Dwivedi, C.; Chaudhary, A.; Srinivasan, S.; Nandi, C.K. Polymer stabilized bimetallic alloy nanoparticles: Synthesis and catalytic application. *Colloid Interface Sci. Commun.* **2018**, *24*, 62–67. [CrossRef]
21. Khlebtsov, B.N.; Liu, Z.; Ye, J.; Khlebtsov, N.G. Au@Ag core/shell cuboids and dumbbells: Optical properties and SERS response. *J. Quant. Spectrosc. Radiat. Transf.* **2015**, *167*, 64–75. [CrossRef]
22. Sangpour, P.; Akhavan, O.; Moshfegh, A.Z. The effect of Au/Ag ratios on surface composition and optical properties of co-sputtered alloy nanoparticles in Au-Ag:SiO$_2$ thin films. *J. Alloy. Compd.* **2009**, *486*, 22–28. [CrossRef]
23. Sangpour, P.; Akhavan, O.; Moshfegh, A.Z. rf reactive co-sputtered Au-Ag alloy nanoparticles in SiO$_2$ thin films. *Appl. Surf. Sci.* **2007**, *253*, 7438–7442. [CrossRef]
24. Hareesh, K.; Joshi, R.P.; Sunitha, D.V.; Bhoraskar, V.N.; Dhole, S.D. Anchoring of Ag-Au alloy nanoparticles on reduced graphene oxide sheets for the reduction of 4-nitrophenol. *Appl. Surf. Sci.* **2016**, *389*, 1050–1055.
25. Tiunov, I.A.; Gorbachevskyy, M.V.; Kopitsyn, D.S.; Kotelev, M.S.; Ivanov, E.V.; Vinokurov, V.A.; Novikov, A.A. Synthesis of large uniform gold and core–shell gold–silver nanoparticles: Effect of temperature control. *Russ. J. Phys. Chem. A* **2016**, *90*, 152–157. [CrossRef]
26. Song, Y.; Liu, K.; Chen, S. AgAu bimetallic janus nanoparticles and their electrocatalytic activity for oxygen reduction in alkaline media. *Langmuir* **2012**, *28*, 17143–17152. [CrossRef] [PubMed]
27. Blaber, M.G.; Arnold, M.D.; Harris, N.; Ford, M.J.; Cortie, M.B. Plasmon absorption in nanospheres: A comparison of sodium, potassium, aluminium, silver and gold. *Phys. B Condens. Matter* **2007**, *394*, 184–187. [CrossRef]
28. Proença, M.; Borges, J.; Rodrigues, M.S.; Meira, D.I.; Sampaio, P.; Dias, J.P.; Pedrosa, P.; Martin, N.; Bundaleski, N.; Teodoro, O.M.N.D.; et al. Nanocomposite thin films based on Au-Ag nanoparticles embedded in a CuO matrix for localized surface plasmon resonance sensing. *Appl. Surf. Sci.* **2019**, *484*, 152–168. [CrossRef]
29. Honda, M.; Ichikawa, Y.; Rozhin, A.G.; Kulinich, S.A. UV plasmonic device for sensing ethanol and acetone. *Appl. Phys. Express* **2018**, *11*, 012001. [CrossRef]
30. Kreno, L.E.; Hupp, J.T.; Van Duyne, R.P. Metal-organic framework thin film for enhanced localized surface Plasmon resonance gas sensing. *Anal. Chem.* **2010**, *82*, 8042–8046. [CrossRef]
31. Sagle, L.B.; Ruvuna, L.K.; Ruemmele, J.A.; Van Duyne, R.P. Advances in localized surface plasmon resonance spectroscopy biosensing. *Nanomedicine* **2011**, *6*, 1447–1462. [CrossRef] [PubMed]
32. Hammond, J.L.; Bhalla, N.; Rafiee, S.D.; Estrela, P. Localized surface plasmon resonance as a biosensing platform for developing countries. *Biosensors* **2014**, *4*, 172–188. [CrossRef] [PubMed]
33. Haes, A.J.; Zou, S.; Schatz, G.C.; Van Duyne, R.P. Nanoscale optical biosensor: Short range distance dependence of the localized surface plasmon resonance of noble metal nanoparticles. *J. Phys. Chem. B* **2004**, *108*, 6961–6968. [CrossRef]
34. Bingham, J.M.; Anker, J.N.; Kreno, L.E.; Duyne, R.P. Van gas sensing with high-resolution localized surface plasmon resonance spectroscopy. *J. Am. Chem. Soc.* **2010**, *132*, 17358–17359. [CrossRef]

35. Demirdjian, B.; Bedu, F.; Ranguis, A.; Ozerov, I.; Henry, C.R. Water adsorption by a sensitive calibrated gold plasmonic nanosensor. *Langmuir* **2018**, *34*, 5381–5385. [CrossRef]
36. Jeong, H.H.; Mark, A.G.; Alarcón-Correa, M.; Kim, I.; Oswald, P.; Lee, T.C.; Fischer, P. Dispersion and shape engineered plasmonic nanosensors. *Nat. Commun.* **2016**, *7*, 11331. [CrossRef] [PubMed]
37. Chen, P.; Liedberg, B. Curvature of the localized surface plasmon resonance peak. *Anal. Chem.* **2014**, *86*, 7399–7405. [CrossRef]
38. Kedem, O.; Vaskevich, A.; Rubinstein, I. Critical issues in localized plasmon sensing. *J. Phys. Chem. C* **2014**, *118*, 8227–8244. [CrossRef]
39. Jung, L.S.; Campbell, C.T.; Chinowsky, T.M.; Mar, M.N.; Yee, S.S. Quantitative interpretation of the response of surface plasmon resonance sensors to adsorbed films. *Langmuir* **1998**, *14*, 5636–5648. [CrossRef]
40. Das, S.; Alford, T.L. Structural and optical properties of Ag-doped copper oxide thin films on polyethylene napthalate substrate prepared by low temperature microwave annealing. *J. Appl. Phys.* **2013**, *113*, 244905. [CrossRef]
41. Rydosz, A.; Szkudlarek, A. Gas-sensing performance of M-doped CuO-based thin films working at different temperatures upon exposure to propane. *Sensors* **2015**, *15*, 20069–20085. [CrossRef]
42. Pedrosa, P.; Fiedler, P.; Lopes, C.; Alves, E.; Barradas, N.P.; Haueisen, J.; Machado, A.V.; Fonseca, C.; Vaz, F. Ag:TiN-coated polyurethane for dry biopotential electrodes: From polymer plasma interface activation to the first EEG measurements. *Plasma Process. Polym.* **2016**, *13*, 341–354. [CrossRef]
43. Barradas, N.P.; Jeynes, C. Advanced physics and algorithms in the IBA DataFurnace. *Nucl. Instrum. Methods Phys. Res. Sect. B* **2008**, *266*, 1875–1879. [CrossRef]
44. Liu, Y.; Zhang, J.; Zhang, W.; Liang, W.; Yu, B.; Xue, J. Effects of annealing temperature on the properties of copper films prepared by magnetron sputtering. *J. Wuhan Univ. Technol. Mater. Sci. Ed.* **2015**, *30*, 92–96. [CrossRef]
45. Pierson, J.F.; Wiederkehr, D.; Billard, A. Reactive magnetron sputtering of copper, silver, and gold. *Thin Solid Films* **2005**, *478*, 196–205. [CrossRef]
46. Figueiredo, V.; Elangovan, E.; Gonçalves, G.; Barquinha, P.; Pereira, L.; Franco, N.; Alves, E.; Martins, R.; Fortunato, E. Effect of post-annealing on the properties of copper oxide thin films obtained from the oxidation of evaporated metallic copper. *Appl. Surf. Sci.* **2008**, *254*, 3949–3954. [CrossRef]
47. Borensztein, Y.; Delannoy, L.; Djedidi, A.; Barrera, R.G.; Louis, C. Monitoring of the plasmon resonance of gold nanoparticles in Au/TiO$_2$ catalyst under oxidative and reducing atmospheres. *J. Phys. Chem. C* **2010**, *114*, 9008–9021. [CrossRef]

© 2019 by the authors. Licensee MDPI, Basel, Switzerland. This article is an open access article distributed under the terms and conditions of the Creative Commons Attribution (CC BY) license (http://creativecommons.org/licenses/by/4.0/).

Article

Nickel Film Deposition with Varying RF Power for the Reduction of Contact Resistance in NiSi

Sunil Babu Eadi [1], Hyeong-Sub Song [1], Hyun-Dong Song [1], Jungwoo Oh [2] and Hi-Deok Lee [1],*

[1] Department of Electronics Engineering, Chungnam National University, Daejeon 34134, Korea; sunil@cnu.ac.kr (S.B.E.); hss2310@cnu.ac.kr (H.-S.S.); hd.song@cnu.ac.kr (H.-D.S.)
[2] School of Integrated Technology, Yonsei Institute of Convergence Technology, Yonsei University, Incheon 21983, Korea; jungwoo.oh@yonsei.ac.kr
* Correspondence: hdlee@cnu.ac.kr; Tel.: +82-042-821-7702

Received: 10 May 2019; Accepted: 27 May 2019; Published: 28 May 2019

Abstract: In this study, the effect of radio frequency (RF) power on nickel (Ni) film deposition was studied to investigate the applications of lowering the contact resistance in the NiSi/Si junction. The RF powers of 100, 150, and 200 W were used for the deposition of the Ni film on an n/p silicon substrate. RMS roughnesses of 1.354, 1.174 and 1.338 nm were obtained at 100, 150, and 200 W, respectively. A circular transmission line model (CTLM) pattern was used to obtain the contact resistance for three different RF-power-deposited films. The lowest contact resistivity of 5.84×10^{-5} Ω-cm^2 was obtained for the NiSi/n-Si substrate for Ni film deposited at 150 W RF power.

Keywords: deposition rate; contact resistance; nickel silicide; radio frequency

1. Introduction

Nickel silicide (NiSi) is a promising metal silicide material for the fabricating source/drain (S/D) contacts in electronic devices; the downscaling of a device leads to an uncontrollable increase in the contact resistance in the S/D and gate electrodes [1–3]. NiSi, by virtue of its characteristic properties such as its low-temperature processing, low silicon consumption, and low resistivity phase compared to other metal silicides, has been studied intensively by various research groups. Ramly et al. reported a study on the controlled diffusion of Ni in the formation of NiSi with different Ni thicknesses for the application of a supercapacitor electrode [4]. Vijselaar et al. studied the effect of a NiSi interlayer on Si substrates for the fabrication of a photoelectrode for photocatalytic properties [5]. Marshall et al. reported the NiSi as a passivated tunneling contact for application in high-efficiency solar cells [6]. However, obtaining a low-resistivity NiSi phase still remains a critical issue for high-efficiency electronic devices. In this context, obtaining low-resistance NiSi is essential for high-performance devices. Previous reports have shown that, by controlling the Ni diffusion through Si, the NiSi phase can be selectively obtained. Kousseifi et al. used a Pt intermixed layer to control the NiSi phase [7]. Jung et al. used an ultraviolet laser to obtain NiSi through a photo-thermal process [8]. Fouet et al. studied the silicide formation using different Ni film thicknesses to control NiSi formation [9]. Tous et al. obtained the direct formation of the NiSi phase using the excimer laser annealing (ELA) process [10]. Different methods have been reported for growing NiSi films for various applications. Kwang et al. studied the interfacial properties of NiSi films deposited by using atomic layer deposition [11]. Mahdi et al. reported well-aligned NiSi/SiC core-shell nanowire growth by hot-wire chemical vapor deposition to enhance the electrical properties of NiSi [12]. Koichi et al. introduced a cyclic deposition process using molecular beam epitaxy (MBE) to grow NiSi for low-resistance films [13]. Azimirad et al. studied the thermal stability of NiSi film by a co-sputtering process and reported the thermal stability of NiSi to be improved by using a platinum interlayer structure on the Si substrate [14].

Therefore, greater understanding of the formation and control of NiSi is needed. To achieve this, the quality and stability of the Ni film are also very important. Ni films can be deposited by various methods such as chemical vapor deposition (CVD), atomic layer deposition (ALD), sputtering, MBE, thermal evaporation, pulsed laser deposition, electroplating [15–21], etc. Among these techniques, vapor deposition through conventional sputtering still remains the most widely used process for metal film deposition. However, to date, the effect of RF power on the contact resistance of the NiSi/Si junctions has not been reported.

In this study, Ni films were deposited with different RF powers, and the influence of the RF power on its surface and structural properties was investigated. Finally, NiSi was obtained by the rapid temperature annealing (RTA) of the different Ni films, and the contact resistance was measured using the circular transmission line model (CTLM) procedure.

2. Materials and Methods

2.1. Ni Films Deposition

The Ni film samples were deposited by using an RF sputtering process on Si (100) substrates. Initially, the substrates were cleaned in dilute HF solution for 150 s and subsequently rinsed in de-ionized water and dried using nitrogen purging. Then, the samples were inserted into the sputtering chamber until the base pressure of the sputtering chamber reached 5.0×10^{-7} Torr. After that, an argon flow of 1.8 sccm flowed into the chamber, and the chamber pressure was maintained at 2.5 mTorr using a pressure gauge. The Ni film was deposited at three different RF powers of 100, 150, and 200 W and the deposition time was fixed to 20 min.

2.2. Nickel Silicide Fabrication and Contact Resistance Measurement

Two types of Si substrates were used to measure the contact resistance of NiSi. Arsenic (As) and boron fluoride (BF$_2$) were used as n and p dopants, by ion implantation with a dose of 5×10^{15} cm^{-2} at 50 keV to obtain n/p-Si substrates, respectively. The in-situ deposition of nickel followed by titanium nitride Ni/TiN [15/10 nm] was performed on n/p-Si substrates at optimized conditions of an Ar flow rate 1.8 sccm and a chamber pressure of 2.5 mTorr and different RF powers of 100, 150 and 200 W, followed by the RTA process at 400 °C for 30 s to obtain NiSi. The TiN film was used as a capping layer to prevent the oxidation of Ni during thermal annealing. The CTLM pattern was fabricated on the substrates to measure the contact resistance. Figure 1 shows the experimental flow chart.

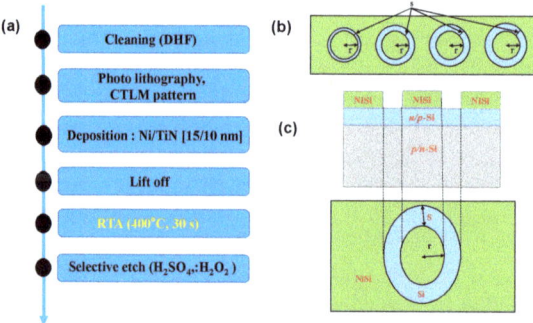

Figure 1. (a) Fabrication flow chart of nickel silicide (NiSi) formations and (b,c) circular transmission line model (CTLM) pattern. r is the radius of the inner circle, and s the radius difference between the inner and outer circle (gap space) and the schematic diagram of NiSi on the Si substrate.

The structural and surface morphologies were characterized using an X-ray diffractometer (XRD, D/MAX 2500PC, Rigaku, Japan) with CuKα radiation, atomic force microscopy (AFM, MAF20 VEECO,

New York, NY, USA), and field-emission scanning electron microscopy (FESEM, Hitachi, S-4800, Tokyo, Japan). The contact resistance was measured using a Kelvin four-point probe (Alessi REL-6100 Cascade Microtech, Beaverton, OR, USA).

3. Results and Discussion

The influence of the RF sputtering power was investigated, keeping all other parameters, such as gas pressure and Ar flow, fixed at 2.5 mTorr and 1.8 sccm, respectively. Figure 2 shows the plot of thickness and resistivity as a function of the sputtering power. The plot result shows that by increasing the RF power, the rate of film deposition increases while the resistivity of the film gradually decreases. The average lowest and highest thicknesses of the deposited Ni films obtained were 61.3 and 100 nm at 100 and 200 W, respectively. The decrease in the resistivity with an increase in the RF power could be due to the decrease in the Ni film crystallite size and surface uniformity. The lowest resistivity of 1.69×10^{-5} Ω-cm was obtained for the 150 W-deposited film. It is possible that as the RF power increased, more Ar ions accelerated to hit the Ni target under high power, resulting in more Ni particles being ejected from the target with increased kinetic energy and velocity, resulting in a faster deposition rate. The high sputtering power improved the crystallization of the Ni films and aided in the formation of the Ni films with highly dense microstructures, which, as a result, led to a reduction in the resistivity [22,23].

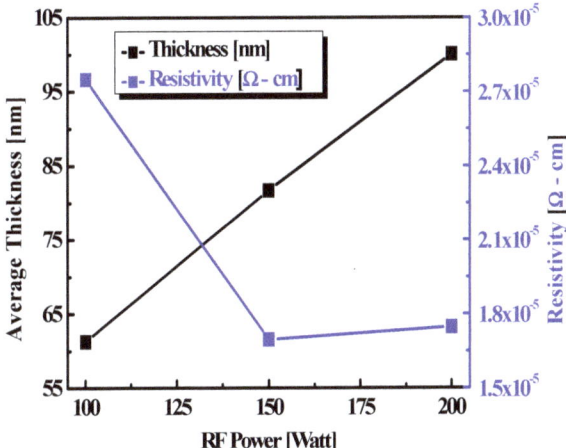

Figure 2. The plot of the Ni films' average thickness and resistivity versus RF power.

To check the surface morphology of the Ni films deposited at various RF powers, FESEM images were taken and are shown in Figure 3. The cross-sectional and surface morphology of the Ni film deposited at 100 W is shown in Figure 3a-1,a-2. The average film thickness of the Ni film was 61.3 nm and the surface shows uniform grain disturbance. However, we notice small cracks on the surface of the Ni film. Further, with an increase in the RF power, the Ni film thickness increased. The average thicknesses of the Ni films were 81.7 and 100 nm for RF powers of 150 and 200 W, respectively, as shown in Figure 3b-2,c-2. The surface morphology of the films shows a bigger crystallite size compared to the 100 W Ni film.

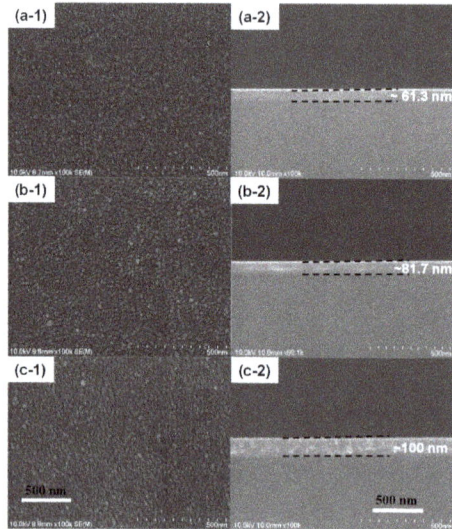

Figure 3. FESEM surface and cross-sectional images of Ni films deposited at different RF powers: 100 W (**a-1,a-2**), 150 W (**b-1,b-2**) and 200 W (**c-1,c-2**). The same scale is used for all images (500 nm).

The structural properties of the films deposited at various magnitudes of RF power are shown in Figure 4. The main peaks at 44.75°, 54.88°, 76.41° are indexed to Ni (111), Ni (200) and Ni (220), which are attributed to Ni's face-centered cubic (FCC) structure [24]. This indicates that the Ni films were in a well-defined crystalline state during deposition. The average crystallite size of the film was determined from the FWHM of the (111) diffraction peaks using Scherrer's equation [25],

$$D_p = \frac{K\lambda}{\beta \cos \theta} \quad (1)$$

where D_p is the average crystallite size, K = Scherrer constant (0.94), λ is the x-ray wavelength (λ = 1.54178 Å), β is the full width half maximum (FWHM) of the (111) peak, and θ is the Bragg angle. It was found that the crystallite size varied between 13.59, 16.24, and 14.91 nm as the RF power changed from 100–200 W. According to the calculations, the crystallite size gradually increased with an increase in the RF power until 150 W, which offered the highest crystallite size, and then gradually decreased. It could be that, with an increase in RF power higher than 150 W, the crystalline quality of film decreases, as evident from the decrease in the relative Ni (111) peak intensity at the higher RF power of 200 W.

To analyze the surface roughness of the Ni films deposited at different RF powers, AFM measurements were obtained. Figure 5 shows the AFM images at different RF powers. The RMS values of the Ni films obtained were 1.354, 1.174, and 1.338 nm. It was noted that the Ni film surface roughness decreases at 150 W and gradually increases at 200 W, indicating that the Ni films deposited at higher RF powers show lower film quality. It is observed that the RF power has an influence on the surface structure of Ni films. The Ni films grown at 150 W RF power have a comparatively low kinetic energy of sputtered particles compared to 200 W RF power, which leads to a relatively more random orientation and various sizes of grain growth, which lead to a rough surface.

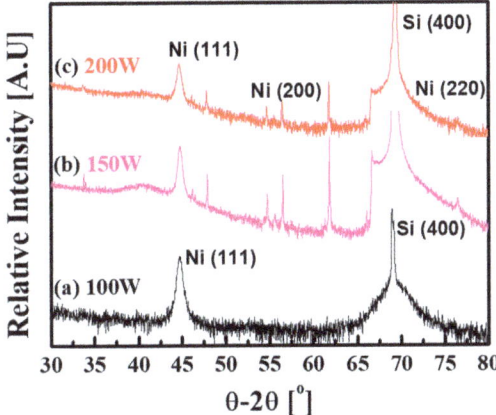

Figure 4. X-Ray diffraction plots of Ni films deposited at different RF powers.

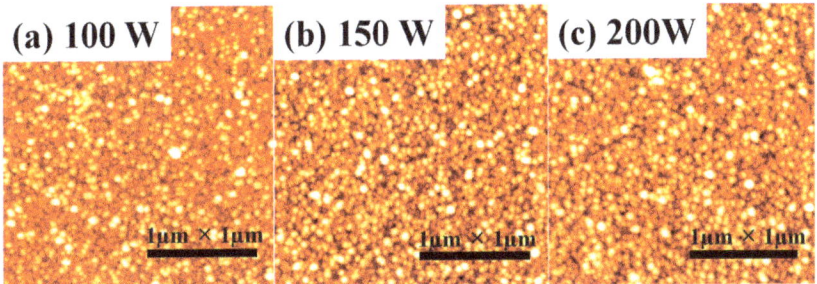

Figure 5. The AFM images Ni films deposited at various RF powers: (**a**) 100 W, (**b**) 150 W and (**c**) 200 W.

To investigate the effect of RF power on the contact resistance, NiSi films were formed with three different RF powers: 100, 150, and 200 W. An Ni film of 15 nm was deposited using different sputtering powers on the Si substrates. Prior to deposition, the deposition rates were calculated for all three RF powers, and the corresponding sheet resistances were measured. The deposition rates were calculated by depositing Ni for different growth times of 11–14 min, and plots of thickness versus growth time and deposition rate were obtained by a linear fit. The deposition rates were found to be 4.6, 5.1, and 7.1 nm/min at 100, 150, and 200 W, respectively.

In the next section, the three RF conditions are referred to as RF-1, RF-2, and RF-3. Initially, to investigate the lowest sheet resistance of the NiSi phase, Ni films deposited on Si substrates were annealed to form NiSi. The RTA process was performed on the samples with varying temperatures from 300–700 °C for 30 s in a N_2 atmosphere. Figure 6 shows the plot of sheet resistance (R_{sh}) verse RTA temperature; it can be observed that Rsh decreases with increasing temperature from 300 °C until 450 °C, and with further increase in temperature, the Rsh increases drastically. It is well known that at a lower temperature (300–400 °C), the NiSi phase predominantly exists, and with an increase in the temperature, Rsh increases. The lowest Rsh values of 6.04 and 6.46 Ω/sq. were obtained for RF-2 on p and n-Si substrates, respectively. Above the RTA temperature of 550 °C, the sheet resistance increases drastically due to the predominant formation of $NiSi_2$, which is the high-resistance phase. Zhao et al. have reported that above 500 °C, the agglomeration of NiSi increases and irregular nucleation of $NiSi_2$ starts, which leads to an increase in resistance [26].

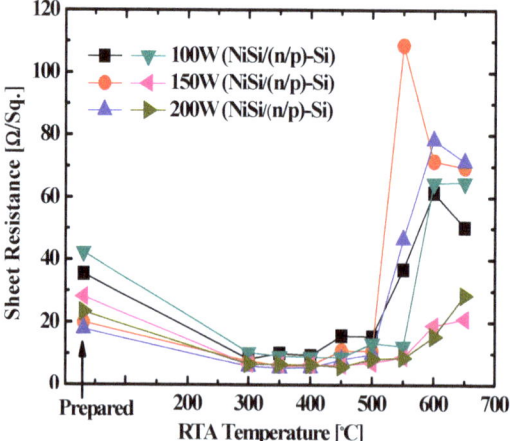

Figure 6. The plot of sheet resistance verses RTA temperature plot of NiSi on n/p-Si formed with different RF powers.

The specific contact resistance ρ_C between the Si substrate and NiSi for the three different Ni films were extracted using a 4-wire Kelvin resistance measurement, and ρ_C was determined using the following equations [27]:

$$R_T = \frac{R_{sh}}{2\pi r}(s + 2L_T)C \qquad (2)$$

$$C = \frac{r}{s}\ln\left(1 + \frac{s}{r}\right) \qquad (3)$$

$$\rho_C = R_{sh}L_T^2 \qquad (4)$$

where R_{sh} is the sheet resistance, C is the correction factor, L_T is the effective transfer length, r is the radius of the inner circle, which was fixed at 80 mm, and s is the gap space, which was split as 8, 12, 16, 20, 24, 32, 40, and 48 mm. R_{sh} and L_T can be determined via a linear fit of R_T at the different gap space values. Figure 7 shows the plot of total resistance measured as a function of the gap space between the inner and outer rings of the NiSi layers for RF-1, RF-2, and RF-3. The lower total contact resistivities of 5.84×10^{-5} Ω-cm^2 for n-Si and 6.58×10^{-5} Ω-cm^2 for p-Si are obtained for the RF-2 film, respectively, and the reduction in contact resistance is lower than the Ni Ohmic contacts previously reported [28]. Thus, by controlling the film deposition using RF power, the quality of the NiSi formation can be controlled. A similar conclusion was reported by Gordillo et al. in their work done on the effect of RF power on electrical properties of Mo films. They concluded from their study that a decrease in the resistivity of Mo films was caused by changing the RF power [29]. In our case, different RF powers control the grain size and surface roughness of the Ni films and show different NiSi electrical characteristics.

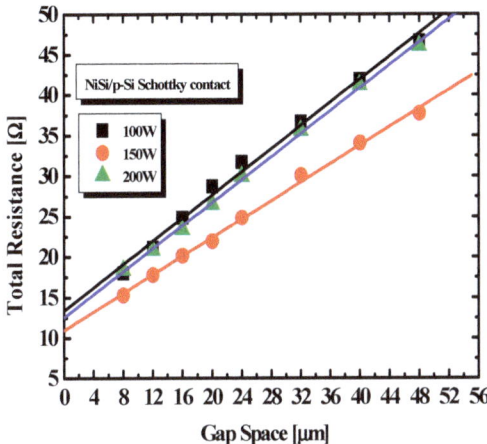

Figure 7. The plot of the total resistance versus gap space of the CTLM pattern for the extraction of specific contact resistivity's with different RF powers.

Table 1 shows the measured values of the contact resistivity and transfer length of the NiSi on n/p-Si substrates. Thus, from the results, it could be that Ni films deposited with an RF power of 150 W showed lower contact resistance in comparison to the Ni films at 100 and 200 W power. It is well known that Ni film and Si substrate undergo a sequential reaction from Ni_2Si to NiSi and finally to $NiSi_2$ with an increase in the RTA temperature [30–32]. In our case, the formation of the sequence of Ni silicide phases of the three films varied with the Ni deposition rate and RTA temperature. The Ni film deposited at 150 W could have controlled the NiSi phase with lower resistivity compared to the other two RF powers. These results lead to the conclusion that, by controlling the Ni film deposition rates, it is possible to obtain low-resistivity NiSi for future electronic devices.

Table 1. The values of the contact resistivity and transfer length of the NiSi on n/p-Si substrates.

Sample	RF Power	NiSi/n-Si		NiSi/p-Si	
		ρ_C [Ω-cm^2]	L_T [μm]	ρ_C [Ω-cm^2]	L_T [μm]
Ni/TiN [15/10 nm]	100 W	8.08×10^{-5}	8.128	7.82×10^{-5}	4.663
	150 W	5.84×10^{-5}	7.493	6.58×10^{-5}	4.787
	200 W	1.18×10^{-4}	10.29	6.94×10^{-5}	4.411

4. Conclusions

In this study, Ni films formed at three different RF powers of 100, 150, and 200 W were investigated for low-resistivity NiSi contacts. The result analysis shows a decrease in the resistivity of Ni films with an increase in RF power. The XRD and AFM data on the Ni films clearly support the change in the resistivity with RF power change. Further, low contact resistance NiSi films were formed by annealing the Ni films, and their contact resistances were measured. The total resistance of 5.84×10^{-5} Ω-cm^2 was obtained for 150 W RF power. The results clearly show that the NiSi formed at 150 W showed lower resistance than NiSi formed at 100 and 200 W. This study is meaningful in that a thin Ni film with a low resistivity was formed by varying the RF power in sputtering, and the formed NiSi phase showed a low contact resistance.

Author Contributions: Conceptualization and Supervision, H.-D.L.; Methodology and Writing—Original Draft Preparation, S.B.E., H.-S.S.; Writing—Review and Editing, H.-D.S., H.-D.L. and J.W.O.; Funding Acquisition, H.-D.L.

Funding: This research was funded by the MOTIE (Ministry of Trade, Industry & Energy (10048536) and the KSRC (Korea Semiconductor Research Consortium) support program for the development of future semiconductor devices. This research was also supported by the Ministry of Trade, Industry and Energy (MOTIE) (10067808) and the Korea Semiconductor Research Consortium (KSRC) support program for the development of future semiconductor devices.

Conflicts of Interest: The authors declare no conflict of interest.

References

1. Rabab, R.B.; Amir, N.H.; Arwa, T.K.; Abdurrahman, G.; Muhammad, M.H. Impact of nickel silicide rear metallization on the series resistance of crystalline silicon solar cells. *Energy Technol.* **2018**, *6*, 1627–1632.
2. Geenen, F.A.; Van Stiphout, K.; Nanakoudis, A.; Bals, S.; Vantomme, A.; Jordan-Sweet, J.; Lavoie, C.; Detavernier, C. Controlling the formation and stability of ultra-thin nickel silicides—An alloying strategy for preventing agglomeration. *J. Appl. Phys.* **2018**, *123*, 075303. [CrossRef]
3. Wolansky, D.; Grabolla, T.; Lenke, T.; Schulze, S.; Zaumseil, P. Impact of nickel silicide on SiGe BiCMOS devices. *Semicond. Sci. Technol.* **2018**, *33*, 124003. [CrossRef]
4. Ramly, M.M.; Omar, F.S.; Rohaizad, A.; Aspanut, Z.; Rahman, S.A.; Goh, B.T. Solid-phase diffusion controlled growth of nickel silicide nanowires for supercapacitor electrode. *Appl. Surf. Sci.* **2018**, *456*, 515–525. [CrossRef]
5. Vijselaar, W.; Tiggelaar, R.M.; Gardeniers, H.J.; Huskens, J. Efficient and stable silicon microwire photocathodes with a nickel silicide interlayer for operation in strongly alkaline solutions. *ACS Energy Lett.* **2018**, *3*, 1086–1092. [CrossRef] [PubMed]
6. Marshall, A.; Florent, K.; Tapriya, A.; Lee, B.G.; Kurinec, S.K.; Young, D.L. Nickel silicide metallization for passivated tunneling contacts for silicon solar cells. In Proceedings of the IEEE 43rd Photovoltaic Specialists Conference (PVSC), Portland, OR, USA, 5–10 June 2016; pp. 2479–2482.
7. El Kousseifi, M.; Hoummada, K.; Bertoglio, M.; Mangelinck, D. Selection of the first Ni silicide phase by controlling the Pt incorporation in the intermixed layer. *Acta Mater.* **2016**, *106*, 193–198. [CrossRef]
8. Jung, S.M.; Kim, J.H.; Park, C.J.; Shin, M.W. Nickel mono-silicide formation using a photo-thermal process assisted by ultra-violet laser. *Mater. Sci. Semicond. Process.* **2018**, *75*, 263–268. [CrossRef]
9. Richard, M.-I.; Mangelinck, D.; Guichet, C.; Thomas, O.; Texier, M.; Fouet, J.; Boudet, N.; Portavoce, A. Silicide formation during reaction between Ni ultra-thin films and Si (001) substrates. *Mater. Lett.* **2013**, *116*, 139–142.
10. Tous, L.; Lerat, J.-F.; Emeraud, T.; Negru, R.; Huet, K.; Uruena, A.; Aleman, M.; Russell, R.; John, J.; Poortmans, J.; et al. Nickel silicide formation using excimer laser annealing. *Energy Procedia* **2012**, *27*, 503–509. [CrossRef]
11. Lee, K.M.; Kim, C.Y.; Choi, C.K.; Yun, S.W.; Ha, J.B.; Lee, J.H.; Lee, J.Y. Interface properties of nickel-silicide films deposited by using plasma-assisted atomic layer deposition. *J. Korean Phys. Soc.* **2009**, *55*, 1153–1157. [CrossRef]
12. Alizadeh, M.; Hamzan, N.B.; Ooi, P.C.; Bin Omar, M.F.; Dee, C.F.; Goh, B.T. Solid-state limited nucleation of NiSi/SiC core-shell nanowires by hot-wire chemical vapor deposition. *Materials* **2019**, *12*, 674. [CrossRef]
13. Terashima, K.; Miura, Y.; Ikarashi, N.; Oshida, M.; Manabe, K.; Yoshihara, T.; Tanaka, M.; Wakabayashi, H. Formation of Nickel Self-Aligned Silicide by Using Cyclic Deposition Method. In Proceedings of the 2004 International Conference on Solid State Devices and Materials, Tokyo, Japan, 14–17 September 2004; pp. 182–183.
14. Azimirad, R.; Kargarian, M.; Akhavan, O.; Moshfegh, A.Z. Improved thermal stability of NiSi nanolayer in Ni-Si Co-sputtered structure. *Int. J. Nanosci. Nanotechnol.* **2011**, *7*, 14–20.
15. Guo, Q.; Guo, Z.; Shi, J.; Sang, L.; Gao, B.; Chen, Q.; Liu, Z.; Wang, X. Fabrication of nickel and nickel carbide thin films by pulsed chemical vapor deposition. *MRS Commun.* **2018**, *8*, 88–94. [CrossRef]
16. Han, W.S.; Kim, S.; Hwang, J.; Park, J.-M.; Koh, W.; Lee, W.-J. Plasma-enhanced atomic layer deposition of nickel thin film using is (1,4-diisopropyl-1,4-diazabutadiene) nickel. *J. Vac. Sci. Technol. A Vacuum Surf. Film* **2017**, *36*, 01A119.
17. Widodo, S. Characterization of Thin Film Nickel (Ni) Deposition by Sputtering Method. *Int. J. Innov. Sci. Technol.* **2015**, *2*, 380–385.

18. Nishiyama, T.; Tanaka, T.; Shikada, K.; Ohtake, M.; Kirino, F.; Futamoto, M. Growth of Ni Thin Films on Al$_2$O$_3$ Single-Crystal Substrates. *Jpn. J. Appl. Phys.* **2009**, *48*, 13003. [CrossRef]
19. Valladares, L.D.L.S.; Ionescu, A.; Holmes, S.; Barnes, C.H.W.; Dominguez, A.B.; Quispe, O.A.; Gonzalez, J.C.; Milana, S.; Barbone, M.; Ferrari, A.C.; et al. Characterization of Ni thin films following thermal oxidation in air. *J. Vac. Sci. Technol. B* **2014**, *32*, 51808. [CrossRef]
20. Kassem, W.; Roumie, M.; Tabbal, M. Pulsed Laser Deposition of Tungsten Thin Films on Graphite. *Adv. Mater.* **2011**, *324*, 77–80. [CrossRef]
21. Li, J.-D.; Zhang, P.; Wu, Y.-H.; Liu, Y.-S.; Xuan, M. Uniformity study of nickel thin-film microstructure deposited by electroplating. *Microsyst. Technol.* **2009**, *15*, 505–510. [CrossRef]
22. Peri, B.; Borah, B.; Dash, R.K. Effect of RF power and gas flow ratio on the growth and morphology of the PECVD SiC thin film s for MEMS applications. *Mater. Sci.* **2015**, *38*, 1105–1112. [CrossRef]
23. Posadowski, W.; Wiatrowski, A.; Kapka, G. Effect of pulsed magnetron sputtering process for the deposition of thin layers of nickel and nickel oxide. *Mater. Sci.* **2018**, *36*, 69–74. [CrossRef]
24. Vergara, J.; Madurga, V. Structure and magnetic properties of Ni films obtained by pulsed laser ablation deposition. *J. Mater.* **2002**, *17*, 2099–2104. [CrossRef]
25. D'Agostino, A.T. Determination of thin metal film thickness by X-ray diffractometry using the Scherrer equation, atomic absorption analysis and transmission/reflection visible spectroscopy. *Anal. Chim. Acta* **1992**, *262*, 269–275. [CrossRef]
26. Zhao, F.F.; Zheng, J.Z.; Shen, Z.X.; Osipowicz, T.; Gao, W.Z.; Chan, L.H. Thermal stability study of NiSi and NiSi$_2$ thin films. *Microelectron. Eng.* **2004**, *71*, 104–111. [CrossRef]
27. Schroder, D.K. *Semiconductor Material and Device Characterization*, 3rd ed.; John Wiley & Sons, Inc.: Hoboken, NJ, USA, 2006; p. 144. ISBN 9780471739067.
28. Kuchuk, A.V.; Borowicz, P.; Wzorek, M.; Borysiewicz, M.; Ratajczak, R.; Golaszewska, K.; Kaminska, E.; Kladko, V.; Piotrowska, A. Ni-based ohmic contacts to n-type 4H-SiC: The formation mechanism and thermal stability. *Adv. Condens. Matter Phys.* **2016**, *2016*, 26. [CrossRef]
29. Gordillo, G.; Mesa, F.; Calderón, C. Electrical and morphological properties of low resistivity Mo thin Films prepared by magnetron sputtering. *Braz. J. Phys.* **2006**, *36*, 982–985. [CrossRef]
30. Zhang, S.-L.; Smith, U. Self-aligned silicides for Ohmic contacts in complementary metal–oxide–semiconductor technology: TiSi$_2$, CoSi$_2$, and NiSi. *J. Vac. Sci. Technol. A Vacuum Surf. Film* **2004**, *22*, 1361–1370. [CrossRef]
31. Chen, X.; Zhang, B.; Li, C.; Shao, Z.; Su, D.; Williams, C.T.; Liang, C. Structural and electrochemical properties of nanostructured nickel silicides by reduction and silicification of high-surface-area nickel oxide. *Mater. Res. Bull.* **2012**, *47*, 867–877. [CrossRef]
32. Lauwers, A.; Kitt, J.A.; Van Dal, M.J.H.; Chamirian, O.; Pawlak, M.A.; De Potter, M.; Lindsay, R.; Raymakers, T.; Pages, X.; Mebarki, B. Ni based suicides for 45 nm CMOS and beyond. *Mater. Sci. Eng. B Solid-State Mater. Adv. Technol.* **2004**, *114–115*, 29–41. [CrossRef]

© 2019 by the authors. Licensee MDPI, Basel, Switzerland. This article is an open access article distributed under the terms and conditions of the Creative Commons Attribution (CC BY) license (http://creativecommons.org/licenses/by/4.0/).

Article

Experimental and Modeling Study of the Fabrication of Mg Nano-Sculpted Films by Magnetron Sputtering Combined with Glancing Angle Deposition

Hui Liang [1,2], Xi Geng [1], Wenjiang Li [1,3,*], Adriano Panepinto [2], Damien Thiry [2], Minfang Chen [1,3,*] and Rony Snyders [2,4,*]

1. School of Materials Science and Engineering, Tianjin University of Technology, Tianjin 300384, China; hui.liang@umons.ac.be (H.L.); xigeng0220@126.com (X.G.)
2. Chimie des Interactions Plasma-Surface, University of Mons, 20 Place du Parc, B 7000 Mons, Belgium; adriano.panepinto@umons.ac.be (A.P.); Damien.THIRY@umons.ac.be (D.T.)
3. Key Laboratory of Display Materials and Photoelectric Device (Ministry of Education), Tianjin 300384, China
4. Materia Nova Research Center, 1 Avenue Nicolas Copernic, B 7000 Mons, Belgium
* Correspondence: liwj@tjut.edu.cn (W.L.); jgj@tjut.edu.cn (M.C.); rony.snyders@umons.ac.be (R.S.); Tel.: +32-65-554955 (R.S.)

Received: 5 May 2019; Accepted: 27 May 2019; Published: 1 June 2019

Abstract: Today, Mg is foreseen as one of the most promising materials for hydrogen storage when prepared as nano-objects. In this context, we have studied the fabrication of Mg nano-sculpted thin films by magnetron sputtering deposition in glancing angle configuration. It is demonstrated that the microstructure of the material is controllable by tuning important deposition parameters such as the tilt angle or the deposition pressure which both strongly affect the shadowing effect during deposition. As an example, the angle formed by the column and the substrate and the intercolumnar space varies between ~20° to ~50° and ~45 to ~120 nm, respectively, when increasing the tilt angle from 60° to 90°. These observations are highlighted by modeling the growth of the material using kinetic Monte Carlo methods which highlights the role of surface diffusion during the synthesis of the coating. This work is a first step towards the development of an air-stable material for hydrogen storage.

Keywords: Mg columnar films; glancing angle deposition; magnetron sputtering; kinetic Monte Carlo modeling

1. Introduction

With increased worldwide energy consumption that is associated with the global warming problem and the depletion of fossil fuels, renewable energy sources from hydro, solar, and wind sources are increasingly replacing the conventional fuels [1,2]. This is the driving force of a real appeal for the development of new solutions in several domains of our society, including the transport industry. Considering the latter, today, several strategies are considered to design the car of the future, and among them, the hydrogen car is one of the most promising ones. Indeed, hydrogen can be produced by various electrochemical and biological methods and has a higher chemical energy as compared with fossil fuels [3–5]. Furthermore, once produced from any energy source, hydrogen generates electricity during fuel cell operations, leaving water vapor as the only exhaust gas, without any other greenhouse gases or harmful emissions [2]. Nevertheless, several issues related to the production, distribution, and storage of hydrogen have to be fixed before using hydrogen as an economically viable fuel for the transport industry [6]. In particular, the hydrogen storage is an important issue related to the low volumetric density of hydrogen. Among the solutions developed to store hydrogen, the utilization of solid-state materials is preferred because of its higher volumetric density (as compared with gaseous and liquid solutions) and for safety reasons. Among the solid-state materials that store hydrogen, the hydride materials where hydrogen is chemically bounded (i.e., not only adsorbed) appear to be good candidates [7–9].

Specifically, magnesium-based hydrides, and more specifically elemental magnesium hydride (MgH_2), are often considered as promising materials for hydrogen storage because magnesium (Mg) is abundant, low cost, has low density, low toxicity and higher hydrogen capacity and reversibility as compared with other hydrides [10,11]. Nevertheless, this material suffers two main drawbacks which are a high desorption temperature and a slow hydrogen sorption kinetic [11]. In addition, Mg can easily be oxidized by oxygen and hydrogen not easily diffused in bulk Mg.

For many years, these problems have been addressed by the community. A complete review on the topic has recently been published by Sadhasivam et al. [12]. From these works, it appears that the reduction of the size of the Mg compounds down to the nanoscale strongly improves the thermodynamic properties of the material [13]. Therefore, several routes have been investigated to reduce the size of the Mg/MgH_2 (below 1 µm) particles such as mechanical ball milling in the presence (or not) of catalyst materials leading to significant improvements in term of the sorption kinetic of the material [14]. Nevertheless, if the sorption kinetic is improved by this approach, this is not the case for the thermodynamic parameters [14]. In order to overcome this problem, it has been suggested that a further reduction of the dimension (<100 nm) of the material could help. This is why efforts have been developed in order to fabricate 1-, 2- or 3D Mg nanoparticles [6,10]. As an example, Barawi et al. [15,16] reported on the synthesis of Mg films by e-beam evaporation on SiO_2 substrates with a thickness ranging from 45 to 900 nm and demonstrated that it plays a major role in the hydrogen absorption kinetics.

In this context and in view of the material science challenges, plasma techniques appear as an ideal technological platform to synthesize these materials. Indeed, these technologies are known as "green" technologies, since they allow for good control of the material properties and their industrial transfer has been demonstrated in many fields such as the glass industry or microelectronics [17–21]. Among these technologies, magnetron sputtering (MS) is usually used to grow dense thin films of various materials (from metal to polymer coatings) [21–23]. Nevertheless, when used in the glancing angle deposition (GLAD) configuration, it has been demonstrated that nanostructurated coatings for the microstructure can be controlled. As an example, we recently reported on the growth of Ti and TiO_2 nanostructurated films by using this approach [19,20].

Therefore, in this work, we aim to study the growth of nano-sculpted Mg films by combining magnetron sputtering and glancing angle deposition (MSGLAD) in order to better understand the growth mechanism of this material which could ultimately be used in composite material for hydrogen storage application. Our strategy consists of a systematic study of the influence of important deposition parameters namely the tilt angle (α) and the working pressure (P_{Tot}) on the microstructure of the synthesized material. These experimental results are compared to computer simulation by Kinetic Monte Carlo (KMC) using the NASCAM code to better understand the growth mechanism of the Mg thin films.

2. Material and Method

2.1. Experimental

All experiments were carried out in a cylindrical stainless-steel chamber (height: 60 cm, diameter: 42 cm), shown in Figure 1. The chamber was evacuated by a turbo-molecular pump (Edwards nEXT400D 160W, Burgess Hill, UK), down to a residual pressure of 10^{-7} Pa. A magnetron cathode was installed at the top of the chamber and the substrate was located at a distance of 80 mm. A 2-inch in diameter and 0.25-inch thick Mg target (99.99% purity) was used. The target was sputtered in DC mode using an Advanced Energy MDK 1.5 K power supply in argon atmosphere using a flow of

12 sccm. Conductive silicon wafers (100) were used as substrates and rinsed with ultra-pure water before deposition.

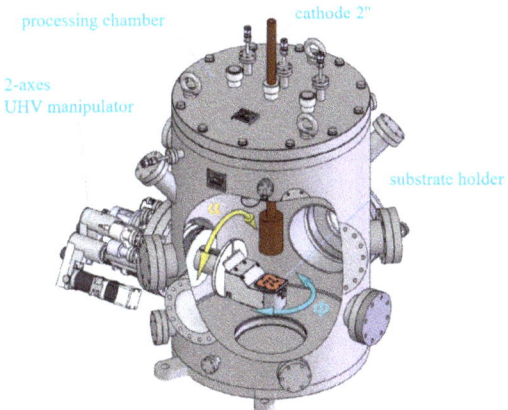

Figure 1. Sketch of the deposition chamber used in this work.

Using the GLAD system, the substrate can be tilted with an angle α and eventually rotated by an angle (φ) either step-by-step or with a continuous angular speed in order to generate diverse thin film architectures. In this work, we have only studied the effect of the tilt angle on the architecture of the deposited films with α = 60°, 80°, 82.5°, 85°, 87°, and 89°. On the other hand, we also evaluated the influence of the working pressure (P_{Tot}) which was varied from 0.13 to 1.3 Pa. For all deposition, the sputtering power was kept constant at 50 W and the deposition time varied between 10 and 20 min depending on the deposition conditions in order to reach similar thicknesses for all deposited films.

The morphology of the material was characterized with a field emission gun scanning electron microscope (FEG-SEM, Hitachi SU8020, Ri Li, Japan). In addition, from the SEM images, we extracted the so-called aspect ratio, Γ, which is defined as the ratio between the inter-columnar space and the column width.

The chemical composition of the films was evaluated by X-ray photoelectron spectroscopy (XPS) on a VERSAPROBE PHI 5000 hemispherical analyzer from Physical Electronics with a base pressure of $<3 \times 10^{-7}$ Pa. The X-ray photoelectron spectra were collected at the take-off angle of 45° with respect to the electron energy analyzer, operating in constant analyzer energy (CAE) mode (23.50 eV). The spectra were recorded with the monochromatic Al Kα radiation (15 kV, 25 W) with a highly focused beam size of 100 μm. The energy resolution was 0.7 eV. Eventual surface charging was compensated for by a built-in electron gun and an argon ion neutralizer. For the chemical depth profile, an Ar^+ ion source was operated at 1 μA and 2 kV with a raster area of 2 mm × 2 mm at an incident angle normal to the sample surface of 54.7°. The XPS spectra were referenced to the Mg2p peak at 49.5 eV arising from the metallic magnesium component [24]. Atomic compositions were derived from peak areas using photoionization cross sections calculated by Scofield, corrected for the dependence of the escape depth on the kinetic energy of the electrons and corrected for the analyzer transmission function of our spectrometer.

The thickness of the films, as measured by a mechanical profilometer Dektak 150 from Veeco, was kept constant for all films, and their average thickness was about 620 ± 20 nm. As an example, the deposition rate was 0.32 nm/s for a deposition angle of 85° and a sputtering pressure of 2 mTorr.

Finally, the phase constitution of the samples was evaluated by X-ray diffraction (XRD) using a PANalytical Empyrean diffractometer working with Cu $K_{\alpha 1}$ radiation (λ = 0.1546 nm) in the grazing incidence configuration (Ω = 0.5°). The X-ray source voltage was fixed at 45 kV and the current at 40 mA. The grain size (G_S) was calculated from the XRD pattern using the following Scherrer equation [25]:

$$G_s = \frac{K\lambda}{\beta \cos\theta}$$

where, K is a dimensionless shape factor, λ is the X-ray wavelength, β is the diffraction line broadening at half the maximum intensity (FWHM), and θ is the Bragg angle.

2.2. Simulation

NASCAM is an atomistic deposition simulation code based on the kinetic Monte Carlo (kMC) method. It can be used for the modeling of different processes occurring at the surface such as the growth of films during deposition. The atoms are deposited on the substrate at random positions at an equal time interval which is determined by the deposition rate. Only diffusion or evaporation events can take place between two deposition events. Energy transfer during ballistic collision events is also taken into account. This made it suitable to simulate glancing angle deposition processes [26]. The energy and angular distribution of incident particles were calculated by SRIM [27] and SIMTRA [28]. First, SRIM was used to calculate the energy and the direction of particles which leave the target. The particles were then transported in the gas phase by the SIMTRA code which took into account all the collisions between the sputtered species and the gas molecules. The energy and the angular distribution of the species at the substrate location were derived for each working conditions by the introduction of the experimental parameters, which included the working pressure, the particles' energy which was a function of the power applied at the target, the racetrack sizes, and the target-to-substrate distance. After that, these files were used as input data for NASCAM. Other parameters could be tuned in the input file. To compare the simulation with the experience, we tuned the number of deposited atoms and the substrate size (XYZ). In these conditions, the simulated and the experimental thin film had the same thickness.

The energy and the angular distribution of the species at the substrate location were derived for each working conditions by the introduction of the experimental parameters such as the working pressure, the power applied to the target, the racetrack size, and the target-to-substrate distance, Other parameters could be varied in the NASCAM input file such as the number of deposited atoms (N) and the substrate size (XYZ). In order to compare simulated and experimental thin films, both had the same thickness. For direct comparison of the cross-sectional film morphology, "2D" (Y = 2) NASCAM simulations were performed (N = 1000 atoms), whereas, for density and porosity evaluation, "3D" (Y = 4) simulations were performed (N = 1.667 atoms). The deposition rate was fixed at 0.5 monolayer by second (0.301 nm/s), which was close to the experimental value (0.32 nm/s) at a deposition angle of 85°.

3. Results and Discussion

3.1. Characterization of A Dense Mg Film

In a first attempt, we have grown a Mg thin film in conventional geometry ($\alpha = 0°$) in order to evaluate the deposition rate, as well as the chemical composition and the phase constitution of the deposited material. Figure 2 shows the survey XPS spectrum recorded for this film. It reveals the presence of Mg, O, and C lines at 49.5 eV (Mg2p), 285 eV (C1s), and 530 eV (O1s). From the quantification of these signal, ~50 atom% of oxygen and 10 atom% of carbon are observed. These are likely related to surface pollution that appears during the transport of the sample from the chamber to the XPS machine. Particularly, the presence of oxygen while working in non-reactive conditions is related strong reactivity of Mg towards O_2 ($\Delta H_{f(MgO)}$ = −601.8 kJ·mol^{-1}) [29], which allows for the oxidation of the top surface of the deposited film. In order to clarify the chemistry of the film and to validate that the presence of carbon, as well as the surface oxidation of Mg, are related to surface pollution, depth profiling of the films were performed by using an Ar^+ gun in the XPS machine before recording the XPS spectra during 2, 4, and 20 min of erosion. The results are presented in Table 1. Clearly, it appears from this analysis that a few minutes of erosion allows removal of all the initially observed carbon contamination as well as reduction of the oxygen content to an ~10 atom% limit, validating the surface contamination of the as-deposited sample. The presence of the 10 atom% of oxygen in the bulk of the material is likely explained by the presence of very low quantity adsorbed water or oxygen in the deposition chamber even if the base pressure is good quality (10^{-7} Torr). Indeed, because of the already mentioned strong reactivity of Mg towards O_2, a getter effect likely occurs and leads to the partial oxidation of the material similar to the effect already observed for another getter material such as Ti [30].

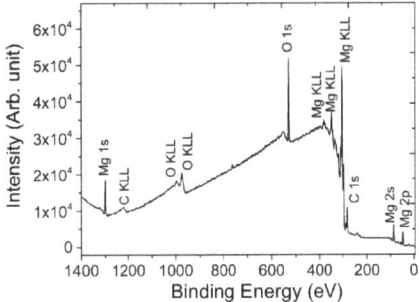

Figure 2. XPS survey spectra of a dense Mg film prepared for $\alpha = 0°$ and $P_{Tot} = 0.26$ Pa. The sputtering power is 50 W.

Table 1. Elemental composition of the as-deposited Mg thin film before and after 2 min of erosion.

At.% Mg		At.% O		At.% C	
As prepared	After erosion	As prepared	After erosion	As prepared	After erosion
44.6	89.9	48.9	10.1	10.5	0

In order to support this conclusion, Figure 3 shows the evolution of the Mg2p XPS line as a function of the depth profiling. The estimated sputtering rate is ~20 nm/min, according to the study reported by Milcius et al. [31]. For the as-deposited sample, it appears that the Mg line is composed of two components corresponding to metallic Mg at 49.5 eV and Mg^{2+} at 50.8 eV. On the other hand, at ~60 eV, a satellite line related to the metallic component is also observable. The presence of a strong oxidized component is in line with the surface stoichiometry of the surface composition of the as-prepared sample. After two minutes of sputtering, which is evaluated to correspond to 40 nm, the oxidized component of the Mg peak completely vanishes while the satellite peak intensity strongly increases. Both these observations clearly confirm that the deposited film is only oxidized on its top surface.

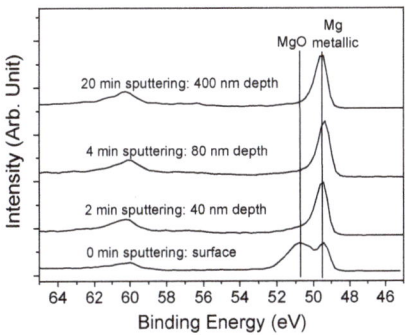

Figure 3. Evolution of the Mg2p line during the depth profiling of the Mg films prepared for $\alpha = 0°$ and $P_{Tot} = 0.26$ Pa.

It has to be noted that since the substrate to target distance (8 cm) is not very high in our chamber, the diameter of the target (2 inches) have to be taken into account to distinguish α and the incident angle of the particles, since the majority of the latter comes from the racetrack region of the target. The size of the particles source induces an angle of deviation in the α direction which increases with the target diameter [33]. This angle of deviation has been calculated in previous work for similar conditions and slightly increases with α due to the geometrical inclination of the substrate leading to an asymmetric deposition. Indeed, the particles sputtered at the left side of the target have a higher probability to reach the left side of the substrate and inversely for the right side. This increases the deviation angle and can explain the similar morphologies for coatings synthesized for α > 85°.

On Figure 6a and Supplementary Figures S1 and S2, the corresponding kMC simulations obtained by using the NASCAM (NAnoSCAle Modeling) code and the procedure described in the experimental part. As input parameters, we have utilized the defined experimental parameters (50 W, 0.26 Pa). A number of 5×10^5 atoms was chosen to obtain a film thickness similar to the one corresponding to the experimental conditions according to the size of the simulation box (X= 1000 and Y = 2 Mg atom unit). The ballistic deposition simulation of Mg atoms was used to understand the growth mechanisms of these films. The morphology of both simulated and experimental thin films were compared and the effect of the deposition parameters was analyzed. From the good agreement between the calculated and experimental data, it appears that the simulation employed in this work is perfectly adapted to our deposition.

Considering that our films are deposited without intentional heating, we can roughly estimate that the deposition temperature is about ~323 K. This corresponds to a T*, the generalized temperature of the Anders's Structure Zone Diagram (ASZM) [17], of ~0.25 since the melting temperature of Mg is 923 K. For this T* condition and considering, in our process conditions, that the normalized energy was < 1 [34], the ASZM depicts the synthesized films as a zone I film corresponding to $T_s/T_m < 0.3$, for which surface diffusion is limited, and therefore does not allow for the filling of the void regions that form in the microstructure because of the geometrical shadowing effect occurring during the GLAD experiments. In these conditions, the film growth proceeds by the formation of an underdense, fine nanofibrous microstructure that develops into a columnar morphology. In the conditions, where geometric restrictions govern the formation of the microstructure, a strong anisotropic deposition is observed. Furthermore, it has to be noted that, for the "zone I" conditions, the columnar tilt angle (β = 44 ± 1.0° in our case) is in line with the Tait's rule derived from geometric analysis of the inter-column shadowing geometry [34].

Figure 6b shows the evolution of β as a function of P_{Tot}. In this work, P_{Tot} was varied from 0.13 to 1.3 Pa where 0.13 Pa corresponds to the minimum value necessary to maintain the magnetron discharge. The sputtering power and α were fixed at 50 W and 85°, respectively. β rapidly decreases as P_{Tot} increases, from β = 51 ± 1.0° for 0.13 Pa to β = 5 ± 0.5° for 1.3 Pa. The modification of the columnar tilt angle can be attributed to a decrease of the collimation of the incident particle flux due to the increase of collision probability as P_{Tot} increases. Indeed, this probability mainly depends on the mean free path of the sputtered Mg atoms (λ_{Mg}) which is inversely proportional to P_{Tot} following this equation:

$$\lambda = \frac{\kappa_B T}{\sqrt{2}\pi d^2 P_{Tot}}$$

where, κ_B is the Boltzmann constant (1.380×10^{-23} J/K), T is the temperature in K, P_{Tot} is the total pressure in Pascal, and d is the diameter of the gas particles in meters. From this relation, λ_{Mg} (atomic diameter = 1.72 Å) ranges from 24 to 2.4 cm between 0.13 and 1.33 Pa, respectively. Considering the target/substrate distance (8 cm) that is used, an increase of P_{Tot} induces a large amount of collisions between particles for the 1.3 Pa conditions resulting in a less porous film. Figure 6b shows that the morphology of the simulated thin films is again in line with the experimental ones. In addition, the calculated mean free path for Ti atoms as a function of the pressure is also presented. It appears that the mean free path becomes smaller than the target-to-substrate distance for a pressure value around

0.7 Pa (λ_{Mg} = 4.5 cm). Below this pressure, very few collisions occur through the vapor phase, whereas, a higher pressure leads to numerous collisions between particles. The analysis of the predicted film morphology at different pressures also allows determination of the range of pressure where a ballistic deposition process occurs which, in our case, is between 0.13 and 0.26 Pa, and its mean free path ranges from 24 cm to 12 cm.

If it is very difficult to measure the porosity of thin films (χ) experimentally because of the low quantity of material. Therefore, it has been simulated using the PoreSTAT software [35] which uses the NASCAM files to perform a full 3D analysis of the porous structure of the material, or a 2D study on the different slices of material belonging to the XZ or YZ planes (X and Y are horizontal axes defining the substrate, whereas, Z corresponds to the vertical axe, defining the film height) [36]. Figure 7 shows the evolution of χ as a function of α for P_{Tot} = 0.26 Pa and of P_{Tot} for α = 85°. It appears that χ increases with α from 54% until 60% for α = 85° and then stabilizes for higher values. On the other hand, χ increases from 51% until 66% when reducing P_{Tot} from 1.33 to 0.13 Pa. These evolutions are obviously correlated with the evolution of the nanosculpted films feature with the α and P_{Tot} parameters. More precisely, it appears from Figure 8 that the evolution of χ is linearly correlated with the evolution of the aspect ratio, Γ, which is convenient since tuning the key parameters of the process such as α or P_{Tot}, we get a fine control on the porosity of the films which is indirectly correlated with the surface area of the material. This linear correlation can be understood by considering the meaning of Γ which is the ratio of the inter columnar space on the width of columns. On the basis of this definition, it is obvious that an increase of Γ will lead to an increase of the material porosity since there is more space between the columns because the intercolumnar space increases or the column width reduces (or both).

The chemical and structural characterization of the nano-sculpted films have again been performed by XPS and XRD measurements. As expected the XPS data reveals the presence of strong oxygen and carbon signals explained by the surface contamination which is likely even stronger for these porous films. Unfortunately, in this case, because of the nanostructured features of the material, it is not easy to depth profile the thin film. To the contrary, XRD measurements are still possible in good conditions and are reported in Figure 2 for a nano-sculpted sample synthesized for α = 85° and P_{Tot} = 0.26 Pa (sputtering power of 50 W).

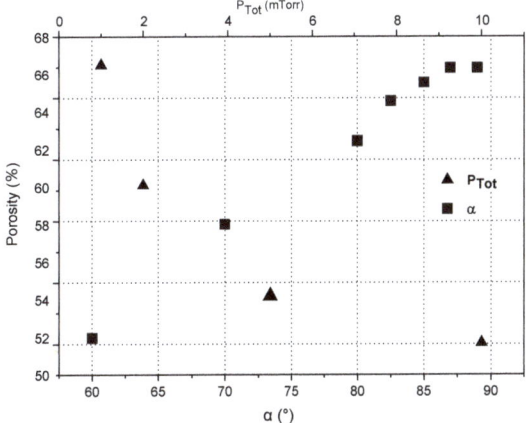

Figure 7. Evolution of the simulated values of the porosity as a function of the tilt angle, α, and of the working pressure, P_{Tot}, for Mg nano-sculpted films prepared using a discharge power of 50 W.

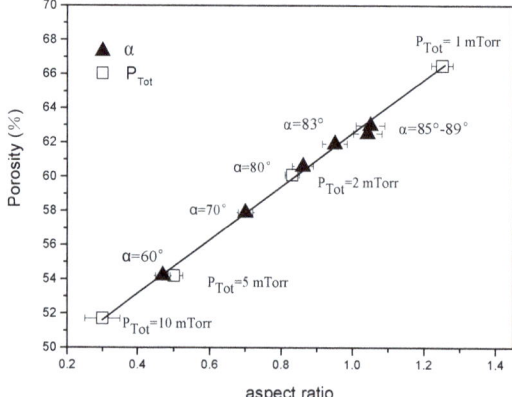

Figure 8. Evolution of the porosity as a function of the aspect ratio, Γ, for Mg nano-sculpted films with the α and P_{Tot}.

From the data we determine that the crystalline constitution of the material is only slightly affected by the utilization of the GLAD geometry. Indeed, all diffraction peaks observed for the dense Mg coatings are again present with the same relative intensity. The only minor difference is related to the presence of MgO lines that appear in addition to the already identified Mg lines. This suggests that the quantity of oxygen in the bulk of the material is likely higher in the nano-sculpted films allowing for the presence of MgO grains. This can be understood when considering the magnified surface area which is subjected to oxidation during the growth as compared with the situation occurring for dense film deposition.

4. Conclusions and Perspectives

Through the present work, we provide a fairly clear description and understanding of a magnetron sputtering in grazing angle geometry method allowing for the deposition of Mg nanocolumnar thin films for potential hydrogen storage. The effect of the deposited angle and sputtering pressure on the Mg nanocolumnar structure has been specifically investigated. The good agreement between experimental observations and model predictions indicates that the simulations realistically reproduces the competitive growth mechanism involved in GLAD experiments. On the basis of this study, we conclude that the fundamental mechanisms responsible for the growth of nano-sculpted Mg film in a MSGLAD are based on (i) the self-shadowing mechanisms at the surface and (ii) the collisional processes of the sputtered particles in the gas phase. In addition, it appears, under our experimental conditions, that the self-diffusion of deposited Mg atoms is strongly reduced and that the microstructure of our films belong to the zone I of the ASZM. In addition, we learn that when growing Mg porous films and because of the strong reactivity of Mg towards oxygen, surface and even bulk oxidation easily occurs. If not controlled, probably, the porous film would not be suitable for hydrogen storage.

Supplementary Materials: The following are available online at http://www.mdpi.com/2079-6412/9/6/361/s1, Figure S1: SEM cross-section view and of Mg films deposited for P_{Tot} = 0.26 Pa and varying α from 60° to 89°. The green images correspond to the structures calculated by using Mkc modeling, Figure S2: SEM cross-section view and of Mg films deposited for α = 85° and varying P_{Tot} from 0.13 to 1.3 Pa. The green images correspond to the structures calculated by using Mkc modeling.

Author Contributions: Conceptualization, H.L. and R.S.; Methodology, H.L.; Software, H.L.; validation, H.L., X.G., A.P. and D.T.; Formal analysis, H.L.; Investigation, H.L.; Resources, H.L.; Data curation, H.L.; Writing—original draft preparation, H.L.; Writing—review and editing, R.S. and W.L.; Visualization, H.L.; Supervision, R.S. and W.L.; Project administration, R.S., W.L. and M.C.; Funding acquisition, W.L. and M.C.; The design of the study, H.L., R.S.; The collection, analyses or interpretation of data, H.L., X.G., A.P. and D.T.; The writing of the manuscript, H.L.; The decision to publish the results, R.S., W.L. and M.C.

Funding: This research was funded by the F.R.I.A grant of the National Fund for Scientific Research (FNRS, Belgium); the Joint Foundation of National Natural Science Foundation of China, Grant No. U1764254; the National Natural Science Foundation of China, Grant No. 51871166; 321 talent projects of Nanjing (China), Grant No. 631783; 111 Project (China), D17003.

Acknowledgments: A.P. thanks the F.R.I.A grant of the National Fund for Scientific Research (FNRS, Belgium); H.L. thanks the Joint Foundation of National Natural Science Foundation of China (Grant No. U1764254); the National Natural Science Foundation of China (Grant No. 51871166); 321 talent projects of Nanjing (Grant No. 631783), China; 111 Project (D17003), China.

Conflicts of Interest: The authors declare no conflict of interest.

References

1. Chaubey, R.; Sahu, S.; James, O.O.; Maity, S. A review on development of industrial processes and emerging techniques for production of hydrogen from renewable an sustainable sources. *Renew. Sustain. Energy Rev.* **2013**, *23*, 443–462. [CrossRef]
2. Momirlan, M.; Veziroglu, T.N. The properties of hydrogen as fuel tomorrow in sustainable energy system for a cleaner planet. *Int. J. Hydrog. Energy* **2005**, *30*, 795–802. [CrossRef]
3. Haas, I.; Gedanken, A. Synthesis of metallic magnesium nanoparticles by sono electrochemistry. *Chem. Commun.* **2008**, *15*, 1795–1797. [CrossRef] [PubMed]
4. Jong, H.D. The preparation of carbon-supported magnesium nanoparticles using melt infiltration. *Chem. Mater.* **2007**, *19*, 6052–6057. [CrossRef]
5. Kooi, B.J.; Palasantzas, G.; De Hosson, J.T.M. Gas-phase synthesis of magnesium nanoparticles: A high-resolution transmission electron microscopy study. *Appl. Phys. Lett.* **2006**, *89*, 161914. [CrossRef]
6. Aneke, M.; Wang, M. Energy storage technologies and real life applications-A state of the art review. *Appl. Energy* **2016**, *179*, 350–377. [CrossRef]
7. Basak, S.; Shashikala, K.; Kulshreshth, S.K. Hydrogen absorption characteristics of Zr sustituted $Ti_{0.85}VFe_{0.15}$ alloy. *Int. J. Hydrog. Energy* **2008**, *33*, 350–355. [CrossRef]
8. Suh, M.P.; Park, H.J.; Prasad, T.K.; Lim, D.W. Hydrogen storage in metal-organic framworks. *Chem. Rev.* **2012**, *112*, 782–835. [CrossRef]
9. Li, W.; Li, C.; Ma, H.; Chen, J. Magnesium nanowires: enhanced kinetics for hydrogen absorption and desorption. *J. Am. Chem. Soc.* **2007**, *129*, 6710–6711. [CrossRef]
10. Sun, Y.H.; Shen, C.Q.; Lai, Q.W.; Liu, W.D.; Wang, W.; Francois, K.; Zinsoua, A. Tailoring magnesium based materials for hydrogen storage through synthesis: Current state of the art. *Energy Storage Mater.* **2017**, *10*, 168–198. [CrossRef]
11. Jain, I.P.; Lal, C.; Jain, A. Hydrogen storage in Mg: A most promising material. *Int. J. Hydrog. Energy* **2010**, *35*, 5133–5144. [CrossRef]
12. Sadhasivam, T.; Kim, H.T.; Jung, S.; Roh, S.H.; Park, J.H.; Jung, H.Y. Dimensional effects of nanostructured Mg/MgH_2 for hydrogen storage applications: A review. *Renew. Sustain. Energy Rev.* **2017**, *72*, 523–534. [CrossRef]
13. Jeon, K.J.; Moon, H.R.; Ruminski, A.M.; Jiang, B.; Kisielowski, C.; Bardhan, R. Air-stable magnesium nanocomposites provide rapid and high-capacity hydrogen storage without using heavy-metal catalysts. *Nat. Mater.* **2011**, *4*, 286–290. [CrossRef] [PubMed]
14. Yuan, J.G.; Zhu, Y.F.; Li, Y.; Li, L.Q. Effect of multi-wall carbon nanotubes supported palladium addition on hydrogen storage properties of magnesium hydride. *Int. J. Hydrog. Energy* **2014**, *39*, 10184–10194. [CrossRef]
15. Barawi, M.; Granero, C.; Chao, P.D.; Manzano, C.V.; Gonzalez, M.; Jimenez, R.D.; Ferrer, I.J.; Ares, J.R.; Fernánde, J.F.; Sanchez, C. Thermal decomposition of noncatalysed MgH_2 film. *Int. J. Hydrog. Energy* **2014**, *39*, 9865–9870. [CrossRef]
16. Dura, J.A.; Kelly, S.T.; Kienzle, P.A.; Her, J.H.; Udovic, T.J.; Majkrzak, C.F.; Chung, C.J.; Clemens, B.M. Porous Mg formation upon dehydrogenation of MgH_2 thin films. *J. Appl. Phys.* **2011**, *109*, 093501. [CrossRef]
17. Laforge, J.M.; Taschuk, M.T.; Brett, M.J. Glancing angle deposition of crystalline zinc oxide nanorods. *Thin Solid Films* **2011**, *519*, 3530–3537. [CrossRef]
18. Thiry, D.; Aparicio, F.J. Surface temperature: A key parameter to control the propanethiol plasma polymer chemistry. *J. Vac. Sci. Technol. A* **2014**, *32*, 050602. [CrossRef]

19. Dervaux, J.; Cormier, P.A.; Konstantinidis, S.; Di, R.; Coulembier, O.; Dubois, P.; Snyders, R. Deposition of porous titanium oxide thin films as anode material for dye sensitized solar cells. *Vacuum* **2015**, *114*, 213–220. [CrossRef]
20. Dervaux, J.; Cormier, P.A.; Konstantinidis, S.; Moskovkin, P.; Lucas, S.; Snyders, R. Nanostructured Ti thin films by combining GLAD and magnetron sputtering and did a joint experimental and modeling study. In Proceedings of the 22nd International Symposium on Plasma Chemistry, Antwerp, Belgium, 5–10 July 2015.
21. Huo, H.W.; Li, Y.; Wang, F.H. Preparation and corrosion resistance of magnesium coatings by magnetron sputtering deposition. *J. Mater. Sci. Technol.* **2003**, *19*, 459–462.
22. Saraiva, M.; Depla, D. Texture and microstructure in co-sputtered Mg-M-O (M = Mg, Al, Cr, Ti, Zr and Y) films. *J. Appl. Phys.* **2012**, *111*, 104903. [CrossRef]
23. Park, C.H.; Lee, W.G.; Kim, D.H.; Ha, H.J.; Ryu, J.Y. Surface discharge characteristics of MgO thin films prepared by RF reactive magnetron sputtering. *Surface Coat. Technol.* **1998**, *110*, 128–135. [CrossRef]
24. Yao, H.B.; Li, Y.; Wee, A.T.S. An XPS investigation of the oxidation/corrosion of melt-spun Mg. *Appl. Surf. Sci.* **2000**, *158*, 112–119. [CrossRef]
25. Scherrer, P. Bestimmung der Grösse und der inneren Struktur von Kolloidteilchen mittels Röntgenstrahlen. *Nachr. Ges. Wiss. Göttingen* **1918**, *26*, 98–100.
26. Lucas, S. Simulation at high temperature of atomic deposition, islands coalescence, Ostwald and inverse ripening with a general simple kinetic Monte Carlo code. *Thin Solid Films* **2010**, *518*, 5355–5361. [CrossRef]
27. Window, B. Recent advances in sputter deposition. *Surf. Coat. Technol.* **1995**, *71*, 91–93. [CrossRef]
28. Zhong, P.; Que, W.; Zhang, J.; Jia, Q.; Wang, W.; Liao, Y. Charge transport and recombination in dye-sensitized solar cells based on hybrid films of TiO_2 particles/TiO_2 nanotube. *J. Alloys Compd.* **2011**, *509*, 7803–7808. [CrossRef]
29. Ruminski, A.M.; Bardhan, R.; Brand, A.; Aloni, S.; Urban, J.J. Synergistic enhancement of hydrogen storage and air stability via Mg nanocrystal–polymer interfacial interactions. *Energy Environ. Sci.* **2013**, *6*, 3267–3271. [CrossRef]
30. Snyders, R.; Gouttebaron, R.; Dauchota, J.P.; Hecq, M. Mass spectrometry diagnostic of dc magnetron reactive sputtering. *J. Anal. At. Spectrom.* **2003**, *18*, 618–623. [CrossRef]
31. Milcius, D.; Grbović-Novaković, J.; Zostautienė, R.; Lelis, M.; Girdzevicius, D.; Urbonavicius, M. Combined XRD and XPS analysis of ex-situ and in-situ plasma hydrogenated magnetron sputtered Mg films. *J. Alloys Compd.* **2015**, *647*, 790–796. [CrossRef]
32. Mor, G.K.; Shankar, K.; Paulose, M.; Varghese, O.K.; Grimes, C.A. Use of highly-ordered TiO_2 nanotube arrays in dye-sensitized solar cells. *Nano Lett.* **2006**, *6*, 215–218. [CrossRef]
33. Pozuelo, M.; Melnyk, C.; Kao, W.H.; Yang, J.M. Cryomilling and spark plasma sintering of nanocrystalline magnesium-based alloy. *J. Mater. Res.* **2011**, *14*, 904–911. [CrossRef]
34. Tait, R.N.; Smy, T.; Brett, M.J. Modelling and characterization of columnar growth in evaporated films. *Thin Solid Films* **1993**, *226*, 196–201. [CrossRef]
35. PoreSTAT. Available online: http://nanoscops.icmse.csic.es/software/porestat (accessed on 31 March 2017).
36. Godinho, V.; Hernández, J.C.; Jamon, D.; Rojas, T.C.; Schierholz, R.; García-López, J.; Ferrer, F.J.; Fernández, A. A new bottom-up methodology to produce silicon layers with a closed porosity nanostructure and reduced refractive index. *Nanotechnology* **2013**, *24*, 275604. [CrossRef]

© 2019 by the authors. Licensee MDPI, Basel, Switzerland. This article is an open access article distributed under the terms and conditions of the Creative Commons Attribution (CC BY) license (http://creativecommons.org/licenses/by/4.0/).

Article

Phase Selectivity in Cr and N Co-Doped TiO$_2$ Films by Modulated Sputter Growth and Post-Deposition Flash-Lamp-Annealing

Raúl Gago [1,*], Slawomir Prucnal [2], René Hübner [2], Frans Munnik [2], David Esteban-Mendoza [1], Ignacio Jiménez [1] and Javier Palomares [1]

[1] Instituto de Ciencia de Materiales de Madrid, Consejo Superior de Investigaciones Científicas, E-28049 Madrid, Spain
[2] Helmholtz-Zentrum Dresden-Rossendorf, Institute of Ion Beam Physics and Materials Research, D-01328 Dresden, Germany
* Correspondence: rgago@icmm.csic.es; Tel.: +34-91-334-9090

Received: 11 June 2019; Accepted: 14 July 2019; Published: 17 July 2019

Abstract: In this paper, we report on the phase selectivity in Cr and N co-doped TiO$_2$ (TiO$_2$:Cr,N) sputtered films by means of interface engineering. In particular, monolithic TiO$_2$:Cr,N films produced by continuous growth conditions result in the formation of a mixed-phase oxide with dominant rutile character. On the contrary, modulated growth by starting with a single-phase anatase TiO$_2$:N buffer layer, can be used to imprint the anatase structure to a subsequent TiO$_2$:Cr,N layer. The robustness of the process with respect to the growth conditions has also been investigated, especially regarding the maximum Cr content (<5 at.%) for single-phase anatase formation. Furthermore, post-deposition flash-lamp-annealing (FLA) in modulated coatings was used to improve the as-grown anatase TiO$_2$:Cr,N phase, as well as to induce dopant activation (N substitutional sites) and diffusion. In this way, Cr can be distributed through the whole film thickness from an initial modulated architecture while preserving the structural phase. Hence, the combination of interface engineering and millisecond-range-FLA opens new opportunities for tailoring the structure of TiO$_2$-based functional materials.

Keywords: oxide materials; doping; sputter deposition; modulated growth; flash-lamp-annealing; XANES

1. Introduction

Titania or titanium dioxide (TiO$_2$) is a functional wide band-gap semiconductor with tuneable electrical and optical properties by intrinsic (structure in single- or mixed-phase anatase/rutile and/or native defects) or extrinsic (doping) mechanisms [1]. The relevance of doping effects comprises many applications that partially rely on the performance of TiO$_2$ as a solvent for impurities. In the case of cation dopants, metal incorporation has been used to functionalize or enhance TiO$_2$ as photocatalyst [2], diluted magnetic semiconductor [3], or transparent conductor material [4].

One of the most interesting properties of TiO$_2$ relies on its photoactivity, which has been exploited in many applications, such as photocatalysis, hydrogen production, pigments or solar cells [2,5]. However, due to the relatively large band-gap of TiO$_2$ (>3 eV), its photoactivity is limited to the ultraviolet (UV) region of the solar spectrum (only 5% of the total energy [6]). Therefore, many efforts have been focused on band-gap narrowing for TiO$_2$ to achieve a visible-light (VISL) response. Such a challenge is mainly realized via doping with foreign atoms at cation or anion sites [2]. In this respect, non-metal (anion) doping has been extensively studied, especially after the work by Asahi et al. [7] on nitrogen (N) doped TiO$_2$ (TiO$_2$:N), where the effective optical absorption appears to be related with intragap localized states [8]. In addition, the solid solubility of N in TiO$_2$ is rather low, and this situation leads to excess N in (a priori, undesirable) interstitial positions [7]. These sites not only compromise the effectiveness of band-gap narrowing, but provide recombination centres responsible for the loss of photo-generated electron-hole pairs [4]. Metal (cation) doping represents another approach to increase VISL absorption in doped TiO$_2$ [9], but it induces structural distortions in the host matrix and the defects act as carrier recombination centres [7]. Among these systems, Cr-doped TiO$_2$ (TiO$_2$:Cr) has been addressed due to its catalytic [10,11] and magnetic [12] properties. A recent concept for effective band-gap narrowing relies on simultaneous doping of TiO$_2$ with anions (C, N, etc.) and transition metals (Cr, V, Mo, etc.). In such a case, the opposite charge states of p- and n-type sites in non-compensated dopants (e.g., Cr/N) should significantly increase the solubility limit of dopant pairs [13]. Based on this hypothesis, Cr–N co-doped TiO$_2$ (TiO$_2$:Cr,N) nanoparticles [6] and single-crystal anatase [14] or rutile [15] thin films have been produced with an extraordinary reduction of the bandgap. The experimental results also indicated that (substitutional) Cr and N dopants are coupled due to the preferential formation of Cr–N bonds [6,14].

In this work, we address the structural impact of co-doping in TiO$_2$:Cr,N films produced by magnetron sputtering. However, Cr containing TiO$_2$ samples produced by this method typically have a poor crystalline quality even under growth on heated substrates up to 500 °C [16]. This eventual drawback represents a handicap with respect to other methods producing TiO$_2$:Cr,N films with high structural quality [13,14]. Post-deposition treatments can be used to enhance the structural quality where, additionally, further dopant activation or the promotion of a specific structure can be achieved. For industry-oriented applications, non-contact and rapid treatments pose a great advantage. Among such methods, millisecond-range flash-lamp-annealing (FLA) [8] enables the control of dopant diffusion and activation, where only the near-surface region is annealed and rapidly cooled. Recently, we reported [17] the structural impact of Cr incorporation before and after FLA in TiO$_2$ films grown by magnetron sputtering. We found that TiO$_2$ phases can accommodate up to ~5 at.% Cr and, in general, rutile environments are favored. This situation may not be desirable for certain applications, since the anatase phase shows a substantially higher photoactivity than rutile [18]. However, phase selectivity in TiO$_2$-based films is a complex competition between nucleation and growth mechanisms. Rutile is the most stable and dense structure, whereas anatase is a metastable phase. The nucleation of rutile requires more energy input than anatase, but after nucleation it grows more easily within a wider range of conditions than anatase [19].

The objectives of this work are two-fold. First, we aim to extend our previous structural study to co-doped films. Second, we seek growth or processing conditions that yield preferential formation of the anatase phase. Since, as stated above, the growth regime for anatase seems to be the limiting factor, we test the preparation of modulated architectures as an attempt to transfer and retain the growth of the anatase phase in the Cr-containing layer. Remarkably, this approach has been found to be rather successful. In addition, FLA has been used to improve the quality of the resulting phase and to induce dopant activation and diffusion. The present results provide a relevant framework for tailoring the atomic structure of TiO$_2$-based mixed oxides by exploiting the concept of interface engineering in combination with FLA.

2. Materials and Methods

2.1. Sample Preparation

In this work, TiO$_2$:Cr,N coatings were grown by reactive magnetron sputtering on p-type (B doped) commercial Si(100) substrates (University Wafer, South Boston, MA, USA) cleaved into 12×12 mm^2 pieces. Individual 3" Ti and Cr targets (99.99% purity, Testbourne Ltd., Basingstoke, UK) were used for co-sputtering, both located at ~15 cm from the grounded substrates in a confocal geometry at an angle of 30° with respect to the substrate normal. The base pressure in the deposition chamber was 10^{-4} Pa. The working pressure was 0.5 Pa with a gas (99.9995% purity grade) mixture of Ar (50%), N$_2$

(44%) and O_2 (6%) set by individual mass flow controllers. For plasma generation, a DC signal with a power of 150 W was applied to the Ti cathode, whereas the power applied to the Cr cathode, W_{Cr}, was modulated from 0 to 25 W according to the desired coating architecture. In all cases, the growth was carried out for 2 h, resulting in a total thickness of ~100 nm. Monolithic and modulated film architectures were studied. In the first case, W_{Cr} was kept constant during the whole deposition time. In the latter case, a buffer TiO_2:N films was initially grown (W_{Cr} = 0 W) with a thickness up to several tens of nm's followed by the subsequent instant (bilayer coating) or gradual (gradient coating) increase in W_{Cr} to produce the Cr-containing layer(s). The deposition was done at a substrate temperature, T_s, of 300 °C to achieve high-quality anatase phase in TiO_2:N films [20].

The as-grown samples were cut in two and one piece was kept as a reference while the other was processed with FLA for 20 ms at a continuous flow of N_2 (99.999% purity). The overall energy density was set to ~70 J/cm^2 according to the optimum conditions for anatase crystallization from (amorphous) pure TiO_2 films [17]. Such FLA condition corresponds to a maximum surface temperature in the range of 1100 °C. The heating and cooling rate during millisecond FLA is in the range of 100 K/ms and 200 K/s, respectively. Further details about the FLA system can be found in [21].

2.2. Sample Characterization

The composition profile of the TiO_2:Cr,N layers was determined by Rutherford backscattering spectrometry (RBS). The measurements were performed at the Ion Beam Center (IBC) of Helmholtz-Zentrum Dresden-Rossendorf (HZDR) using a 1.7 MeV He$^+$ probing beam. The RBS spectra were acquired under normal incidence with a "random" scan (to avoid channelling effects in the substrate signal), and the backscattered particles were detected with a silicon detector at a scattering angle of 165°. For quantitative analysis, the RBS spectra were simulated with the SIMNRA code [22]. Complementary compositional analysis regarding light elements was done by means of heavy-ion elastic recoil detection analysis (ERDA). The measurements were carried out with a 35 MeV Cl^{7+} beam impinging at 75° with respect to the sample surface normal. The scattered ions and recoils were detected with a Bragg ionization chamber (BIC) located at a scattering angle of 31°. The BIC chamber allows for the discrimination of detected particles according to their atomic number (Z). The analysis of the recoil spectra and scattered Cl spectrum was performed simultaneously for each sample with the code NDF [23].

The phase structure of the samples was examined by grazing-incidence X-ray diffraction (GI-XRD) measurements using a D5000 (BRUKER AXS, Billerica, MA, USA) diffractometer with Cu-Kα radiation (wavelength of 1.5418 Å). The data were collected at an incidence angle of 0.5°. The local bonding structure with element sensitivity was studied by X-ray absorption near-edge structure (XANES) analysis [24] with soft X-rays. XANES provides short-range information of electronic states for each individual element, being a powerful technique to study complex multi-element systems. Moreover, the technique can be applied to materials with either amorphous or crystalline structure. Particularly, TiO_2-based materials have been widely studied by XANES, providing clear distinct spectral fingerprints of TiO_2 polymorphs that can be used for univocal phase identification [25]. XANES measurements were carried out at the dipole beamline PM3 of the synchrotron facility BESSY-II of Helmholtz-Zentrum Berlin (HZB). The data were acquired using the ALICE endstation in the total electron yield (TEY) mode. TEY-XANES probes the near-surface region (up to a few tens of nm's) and, therefore, in-depth structural information can be achieved by the combination of XANES and XRD. The electronic structure has also been studied by high-resolution X-ray photoelectron spectroscopy (XPS). The spectra were acquired with a Phoibos 150 spectrometer (SPECS, Berlin, Germany) equipped with a hemispherical analyser and 2D-DLD detector. The measurements were performed with monochromatic Al Kα radiation at normal emission take-off angle. The spectra were acquired using an energy step of 0.05 eV and pass-energy of 10 eV, providing an overall instrumental peak broadening of ~0.4 eV.

The microstructure of the TiO_2:Cr,N thin films was analysed locally with cross-sectional transmission electron microscopy (TEM) using an image C_s-corrected Titan 80–300 microscope

(FEI, Eindhoven, Netherlands) operated at an accelerating voltage of 300 kV. TEM specimens were prepared by sawing, grinding, dimpling, and finally, Ar$^+$ ion-milling. In particular, bright-field TEM imaging and selected-area electron diffraction (SAED) were performed. Since the smallest available selected area aperture of 10 µm covers a circular area with a diameter of about 190 nm, amorphous glue used for TEM specimen preparation contributes to the SAED patterns recorded. Complementary information was extracted from high-angle annular dark-field scanning transmission electron microscopy (HAADF-STEM) imaging and spectrum imaging analysis based on energy-dispersive X-ray spectroscopy (EDXS). This analysis was performed with a Talos F200X microscope (FEI, Brno, Czech Republic) operated at 200 kV and equipped with a Super-X EDXS detector system. The TEM specimens were placed for 8 s into a Model 1020 Plasma Cleaner (Fischione, Export, PA, USA) to remove organic surface contamination before the analysis.

3. Results and Discussion

3.1. Compositional Profile in Monolithic and Modulated Films

Figure 1 shows the experimental (dots) and fitted (solid lines) RBS data from monolithic and modulated TiO$_2$:Cr,N coatings. In this case, the monolithic film was produced with constant power W_{Cr} = 15 W during the whole deposition time. The bilayer coating was produced with W_{Cr} = 0 W for one half of the deposition time (~50 nm), followed by switching on the discharge on the Cr cathode to W_{Cr} = 15 W during the rest of the process. Finally, the gradient coating was produced with W_{Cr} = 0 W for one fourth of the total deposition time (~25 nm) and a stepwise increase of W_{Cr} by 5 W each 30 min. The designed layer structures for the different coating architectures are depicted as inset in Figure 1. The contributions from the various elements to the overall RBS spectra are also labelled in the figure. The Cr contribution to the overall spectra obtained from the fitting results is also included in the graph and reflects the different layer arrangements. Moreover, the Cr doping level is similar (~4 at.%) at the near surface region for all cases (where W_{Cr} = 15 W), with a Cr/Ti ratio of around 0.15. Obviously, the Cr concentration is constant for the whole film thickness in the monolithic film. The RBS cross-section is low for light elements and, hence, only the oxygen signal is clearly detected due to its large concentration. However, the N content is found to be around 2–3 at.%, as derived by complementary ERDA measurements.

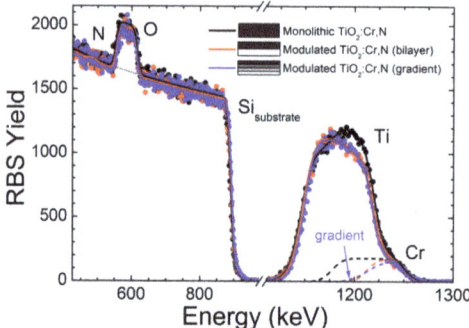

Figure 1. RBS (Rutherford backscattering spectrometry) data (dots) and fitted spectra (solid lines) for as-grown TiO$_2$:Cr,N coatings grown on Si(100) substrates with monolithic and modulated (bilayer and gradient architectures) structures. The individual element contribution from Cr to the fitted spectra is also shown (dotted lines).

3.2. Structural Investigations of Monolithic and Modulated Films

Figure 2 shows the GI-XRD patterns from monolithic and modulated TiO_2:Cr,N coatings as well as from TiO_2 and TiO_2:N monolithic films grown under equivalent conditions. The TiO_2 coating shows a mixture of rutile and anatase phases for this particular growth condition, although the relative phase contribution can be tuned, among other parameters, by the oxygen partial pressure [26]. Note that the broad bump around 55° comes from the underlying Si(100) substrate. The dominant A_{101} reflection from the TiO_2:N coating evidences that the addition of N under these conditions prevents rutile formation and yields a nearly single-phase anatase film. This effect was reported in detail in a previous publication by some of the present authors [20] and other groups [27–29]. In the Cr co-doped monolithic film, the structure displays a strong mixed-phase character where the rutile content is significantly higher than in the pure TiO_2 film. As discussed in the introduction, this result supports the favourable formation of rutile environments upon Cr incorporation. The R_{110} reflection from rutile in the monolithic TiO_2:Cr,N coating is considerably broad, which can be attributed to both the structural disorder induced by Cr incorporation into the oxide matrix and, eventually, the formation of a rutile hybrid oxide in a similar fashion as observed in Cr-doped TiO_2 [17]. Remarkably, the anatase selectivity in the TiO_2:N buffer layer can be effectively used to imprint this structural phase to the co-doped layer in the modulated growth. Indeed, both modulated films display strong A_{101} Bragg reflections corresponding to a nearly single-phase structure. The similar result in both modulated designs supports the validity of the transfer process, where anatase keeps on growing by either increasing the Cr content progressively or abruptly. The widths of the A_{101} Bragg reflection indicate that the anatase phase is nanocrystalline, with grain sizes of around 20 nm, as estimated by the Scherrer formula [30]. Finally, in all Cr-containing films, no evidence of Cr–O phase segregation is observed.

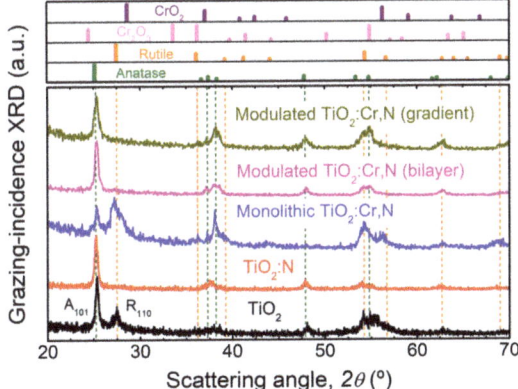

Figure 2. Grazing-incidence XRD patterns from TiO_2, TiO_2:N and TiO_2:Cr,N (monolithic/modulated) coatings grown under equivalent conditions. The diffraction patterns from anatase (PDF-00-021-1272), rutile (PDF-00-021-1276), Cr_2O_3 (PDF-00-001-0622) and CrO_2 (PDF-00-001-0622) reference compounds are shown in the upper panel for phase identification.

For complementary phase identification at the near-surface region, the bonding structure around host and dopant sites has been studied in detail by XANES. The spectra for the Ti 2p, O 1s, Cr 2p and N 1s element edges are shown in Figure 3. For the sake of clarity, only the spectra from the bilayer coating are shown as representative case of the modulated growth (same results were obtained from the gradient design). The element spectra from corresponding reference binary oxides as extracted from the literature are also shown for comparison. In particular, the TiO_2 references refer to the spectra reported for polycrystalline films by Ruus et al. [31]. Regarding chromium oxides, we have included the spectra from single-crystal Cr_2O_3 [32], CrO_2 deposited film [33] and CrO_3 powder [34]. Accordingly, the spectral features from the latter oxides can be used as fingerprints of environments with oxidation state Cr^{3+}, Cr^{4+} and Cr^{6+}, respectively. Basically, the individual XANES spectra can be interpreted as a picture of the density of Ti-$3d$, O-$2p$, N-$2p$, and Cr-$3d$ states according to dipole selection rules. A detailed description of the origin of the different spectral features and regions can be found in previous publications for binary [26,35] and ternary [17,20] oxides. The most direct information about the oxide matrix can be found in the Ti $2p$ edge, where anatase and rutile TiO_2 display clear distinct features, especially in the double-peak structure within the 459–462 eV region. In this way, the Ti $2p$ line shape indicates a dominant rutile and anatase structure for monolithic and modulated films, respectively. Then, surface-sensitive XANES results prove that the anatase phase in the TiO_2:N buffer layer is clearly transferred from the interface up to the surface in the Cr-containing layer by the modulated growth.

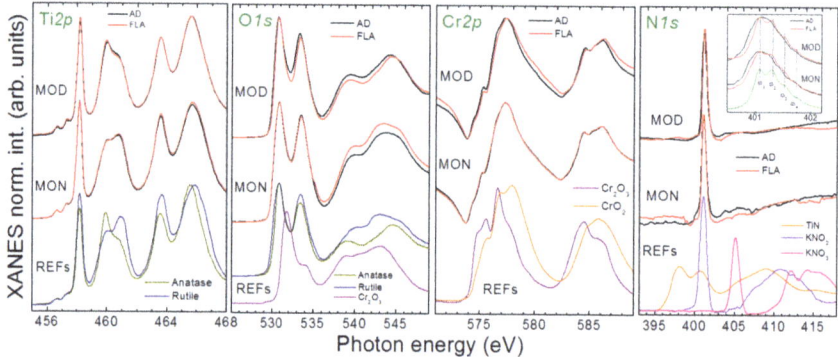

Figure 3. XANES spectra at the different element edges for monolithic (MON) and modulated (MOD) TiO_2:Cr,N coatings before (black) and after (red) FLA. The reference spectra from binary oxide compounds are also included in the bottom part (see text for details).

The O $1s$ leads to similar conclusions in relation to the dominant titania phase. Moreover, there is no evidence of O–Cr bonds, since the signal is dominated by the TiO_2 matrix. In any case, information about the Cr sites can be extracted from the Cr $2p$ edge. The Cr $2p$ spectra are rather broad in comparison to the reference oxides due to the nanocrystalline nature of the films. The spectra show a marked background since the Cr $2p$ edge is superimposed to the O $1s$ post-edge and, in addition, the Cr content in the films is relatively low. Here, the adsorption onset and spectral features indicate that most of the Cr dopants are in the Cr^{3+} oxidation state [17,35]. This situation has been considered as an indication of the substitutional nature of Cr dopants [6] and, from a practical point of view, can also be relevant to improve the photocatalytic response [10,11]. However, a certain contribution from Cr^{4+} (CrO_2) may also be plausible by the increased intensity in the 578–579 and 586–588 eV regions. The presence of Cr^{4+} sites could also be correlated with the formation of a (secondary) mixed-oxide rutile phase [17], as suggested by the rather broad R_{110} Bragg reflection in the GI-XRD pattern from the monolithic TiO_2:Cr,N in Figure 2. Regarding N sites, the XANES N $1s$ shows a main contribution of interstitial sites in the form of –NO_x radicals ($x \sim 2$), as evidenced by the intense peak at ~401 eV. However, as shown for the case of N-doped TiO_2 [26], such environments may have a relatively large cross-section and some N can also be in substitutional sites (N–Ti). This is further discussed below when presenting the XPS results.

3.3. Effect of Flash-Lamp-Annealing (FLA)

The impact of FLA on the structural properties of monolithic and modulated films was also examined. XANES spectra of the flashed (red curves) and as-deposited (black curves) films are compared in Figure 3. Since the films were grown at moderate temperatures, the modifications induced by FLA are not severe. However, there are subtle changes that are worth mentioning. First, the low energy peak at ~461 eV is promoted and sharpened at the Ti 2*p* edge of the modulated film, which evidences a slight increase in the quality of the anatase phase. A similar structural improvement is hinted in the rutile features for the monolithic film. Another relevant change induced by FLA in both structures is related to the nature of N environments. In particular, the high-resolution scan of the N 1*s* peak around 401 eV (see inset) reveals the appearance of a fine structure, a fingerprint of the formation of N_2 molecules inside the oxide matrix [20].

The structural changes induced by FLA were also studied by GI-XRD (note that this analysis was performed in smaller samples than those in Figure 2). In agreement with XANES, the XRD patterns in Figure 4 also show that the phase composition is not significantly altered by FLA. In this case, a small transformation into rutile takes place in the bilayer coating, which is not observed for the case of a gradient profile. In addition, the A_{101} Bragg peak in modulated coatings displays a shift after FLA to higher angles (smaller lattice parameter). This should be mainly attributed to strain effects, but another plausible contribution could be the promotion of N substitutional sites (see below).

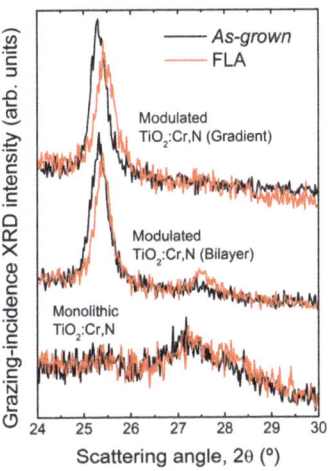

Figure 4. Grazing-incidence XRD patterns from monolithic and modulated TiO_2:Cr,N coatings before and after FLA. The anatase phase in modulated coatings shows a shift to higher scattering angles (lattice contraction) with the thermal treatment.

In order to deepen in the evolution of N environments, XPS was performed in as-deposited and FLA samples. The N 1s core-level spectra for the different samples are shown in Figure 5. In line with the XANES, XPS of as-grown samples shows the dominance of interstitial sites (N–O), together with the presence of N_2 molecules (N–N). Note that FLA increases the relative contribution of N–N bonds, as already pointed out by XANES. Remarkably, FLA also transforms N into substitutional sites (Ti–N) and, hence, it clearly favors activation of the anion dopant.

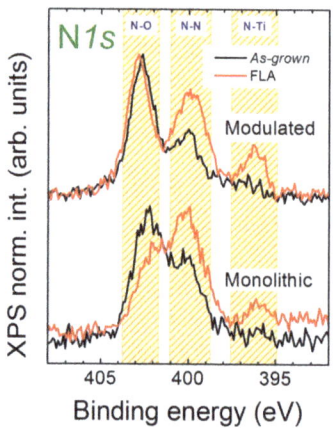

Figure 5. N 1s core-level spectra from monolithic and modulated TiO_2:Cr,N coatings before and after FLA.

The impact of FLA on the film morphology and phase structure was also studied by TEM analysis. Figure 6 shows cross-sectional bright-field TEM images of the monolithic and modulated architecture with a gradient profile. In both samples, there is a thin (~2 nm) SiO_2 layer at the substrate-coating interface coming from the native oxide of the Si(100) wafers. The images show columnar grain growth with in-plane diameters of a few tens of nanometres, which is consistent with the grain size estimated by GI-XRD. Irrespective of the film architecture, the crystal growth is more or less homogenous across the whole layer thickness. The samples after FLA show a more pronounced grain orientation contrast close to the interface to the substrate, which can be correlated with an enhanced crystallinity in that region. The presence of pores in the early stage of growth can also be seen in both samples, although they are more evident in the TiO_2:N buffer layer within the modulated design. Such pore structures can be resolved more clearly in slightly under-focused (UF) bright-field TEM images, as shown in Figure 6 from a magnified view of the film-substrate interface region. It is also relevant that those structures increase in size after FLA. The UF images also emphasize the enhancement (darker zones) of the film crystalline quality near the substrate interface after FLA. Finally, selected-area electron diffraction (SAED) analysis shown in Figure 6 from the imaged regions supports the phase formation derived from XANES and XRD, with a dominant rutile or anatase character for monolithic and modulated coatings, respectively.

Spectrum imaging analysis based on EDXS was done at the cross-sectional TEM specimens to study the in-depth element distributions. The corresponding HAADF-STEM micrographs and the EDXS element maps are shown in Figure 7. The images reveal homogenous distributions of Ti, O and N atoms. Note that there is also an additional narrow N contribution below the native oxide layer, which should be attributed to the commercial wafer fabrication process. The Cr distributions from the as-deposited samples clearly reproduce the coating designs, where the different interfaces resulting from the modulated growth can be identified, as highlighted by the dashed lines in the figure. The preservation of the expected profile also evidences that TEM specimen preparation has not altered the element distributions. Interestingly, the Cr distribution obtained after FLA in the modulated structure shows that the thermal treatment can (at least partially) homogenize the initial Cr profile. Therefore, one could design a modulated structure for phase selectivity and subsequently distribute the Cr dopants by FLA. Quantitative EDXS analysis (line scans not shown here) provide similar results as RBS and ERD with Cr and N contents of a few at.%. In the monolithic film, there is a slight accumulation of N in the region where the aforementioned pores are observed (note the contrast in this region in the HAADF-STEM images due to the reduced atomic number). Therefore, we assume that such mesoporous structure may be filled with N_2 (bubbles). Note that the promotion of N_2 after FLA has been detected at the near-surface region by XANES and XPS. In this regard, it should be mentioned that the reduced N signal in the pore region for the FLA-treated modulated film (see Figure 7) is most probably caused by N_2 release from the enlarged and interconnected pores, which get opened during preparation of the thin electron-transparent TEM lamella.

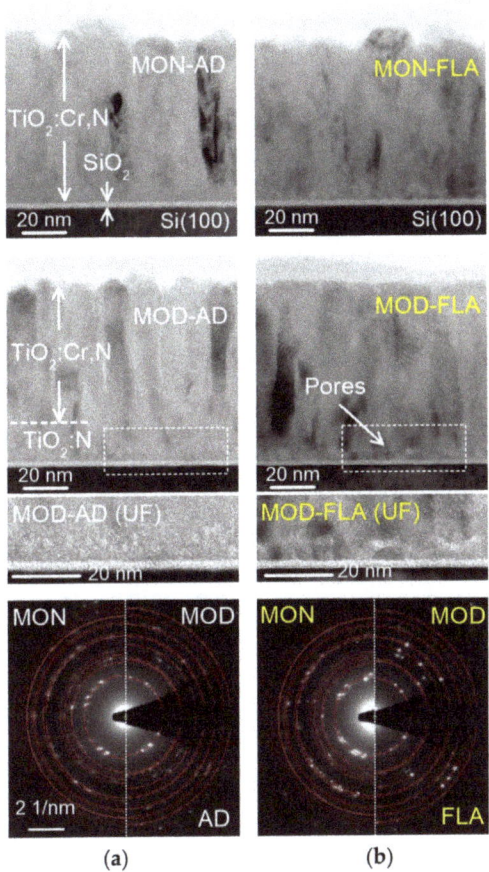

Figure 6. Cross-sectional bright-field TEM images from monolithic (MON) and modulated (MOD) TiO_2:Cr,N coatings as-deposited (AD) (**a**) and after FLA (**b**). The slightly under-focused (UF) bright-field images in the third row are magnified views of the marked regions close to the substrate-coating interface in the MOD case to enhance the contrast of the pore structures. Corresponding representative SAED patterns displayed in the bottom part confirm the dominant rutile and anatase character for MON and MOD coatings, respectively.

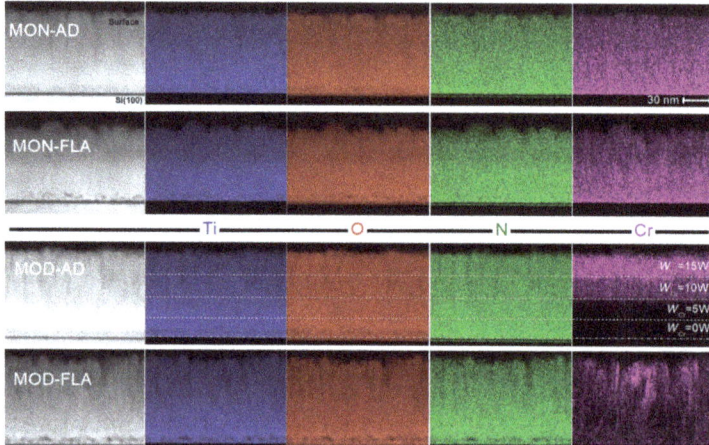

Figure 7. HAADF-STEM micrographs and element distributions obtained by EDXS analysis of monolithic (MON) and modulated (MOD) coatings in as-deposited (AD) and after FLA states. Dashed lines in the MOD-AD coating indicate the interfaces caused by the gradient steps during film growth.

3.4. Final Remarks

One relevant question regarding the modulated growth relies on the generalization of the phase selectivity for other growth conditions and coating designs. For this purpose, additional processing windows were tested. First, higher W_{Cr} were sampled for the stability of the anatase phase against higher Cr contents in the films. In particular, bilayer films were produced with W_{Cr} up to 25 W in the second growth stage. This change implies a slight increase of the Cr content from ~4 to ~5 at.%, as extracted from RBS. As shown in the Ti 2p XANES spectra in Figure 8, the progressive increase in Cr results in a higher promotion of the rutile phase. Hence, single-phase anatase TiO$_2$:Cr,N layers produced in modulated structures can only accommodate Cr contents below the threshold of ~5 at.%. Interestingly, the Ti 2p spectra from the samples after FLA (red curves) confirm that the thermal treatment improves the crystal structure, and even restores the anatase dominance for the highest Cr contents. The latter could be related to the aforementioned redistribution of Cr after FLA, which would imply an effective decrease in the Cr content. Such a hypothesis is supported by the relative intensity of the corresponding XANES edges (not shown).

Another crucial parameter in the modulated growth is the substrate temperature. As indicated in the experimental section, we selected the lowest temperature (300 °C) capable of producing high-quality anatase TiO$_2$:N layer [20] for the buffer layer. This means that we could obtain better (worse) quality anatase phase at higher (lower) temperatures. The thickness of the Cr-containing layer should also be considered to test how long the anatase growth can be sustained in the modulated mode. To study this influence, we prepared bilayer coatings by increasing the Cr-containing layer from 50 to 75 nm. In the latter case, the structure preserves the anatase character, but the rutile contribution starts to increase for the thicker layer.

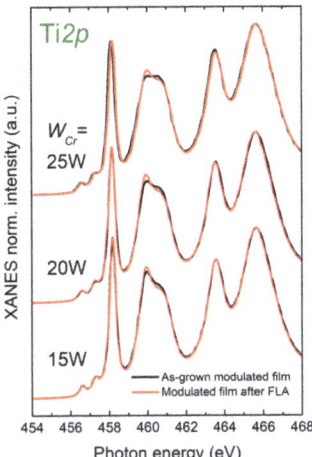

Figure 8. XANES spectra of modulated films with a bilayer structure produced with different W_{Cr} in the uppermost layer in as-grown (black curves) and after FLA (red curves) states.

Finally, in the present work, we have focused our investigations on the structural properties related to the modulated growth. Obviously, additional efforts should be devoted to varying the FLA conditions and modulated designs in order to optimize the structural quality. In addition, a natural continuation of this work would be testing the performance of these coatings for photocatalytic applications.

4. Conclusions

In conclusion, we have studied the phase formation in TiO_2:Cr,N thin films and exploited interface engineering concepts for phase selectivity. In particular, monolithic coatings result in the formation of a mixed-phase oxide with dominant rutile character. Remarkably, under equivalent conditions, the favourable formation of single-phase anatase in a TiO_2:N buffer layer can be utilized to transfer this structure into TiO_2:Cr,N. Such phase selectivity in the Cr-containing film occurs in both gradual and abrupt Cr profiles. However, the production of single-phase anatase seems to be limited to low Cr contents (<5 at.%) and is less effective as the film thickness is increased. Post-deposition millisecond-range FLA was used to enhance the quality of the as-grown phase together with dopant activation and diffusion. This work shows that the combination of modulated growth designs and FLA opens new opportunities for tailoring the desired structure of TiO_2-based materials.

Author Contributions: Conceptualization, R.G. and S.P.; Methodology, R.G. and S.P.; Validation, R.G., S.P., and R.H.; Formal Analysis, R.G., S.P., R.H. and F.M.; Investigation, R.G., S.P., R.H., F.M., D.E.-M., I.J. and J.P.; Resources, R.G., S.P., R.H., F.M. and J.P.; Writing—Original Draft Preparation, R.G.; Writing—Review and Editing, R.G., S.P., R.H., F.M., D.E.-M., I.J. and J.P.; Visualization, R.G. and R.H.; Supervision, R.G. and S.P.; Project Administration, R.G. and S.P.; Funding Acquisition, R.G., S.P., R.H., I.J., and J.P.

Funding: The research leading to these results has received funding from projects RTI2018-095137-B-I00 and MAT2016-80394-R from *Ministerio de Ciencia, Innovación y Universidades* (Spain) and grant agreement n° 226716 from the European Community's Seventh Framework Programme (FP7/2007-2013). Funding of TEM Talos F200X by the German Federal Ministry of Education of Research (BMBF) through Grant No. 03SF0451 in the framework of HEMCP is gratefully acknowledged.

Acknowledgments: We thank HZB and HZDR for the allocation of beamtime for the synchrotron and ion beam analysis experiments, respectively. Support by the TEM facilities at IBC-HZDR is gratefully acknowledged. The authors would also like to thank R. Aniol at HZDR for TEM specimen preparation.

Conflicts of Interest: The authors declare no conflict of interest.

References

1. Diebold, U. The surface science of titanium dioxide. *Surf. Sci. Rep.* **2003**, *48*, 53–229. [CrossRef]
2. Henderson, M.A. A surface science perspective on TiO_2 photocatalysis. *Surf. Sci. Rep.* **2011**, *66*, 185–297. [CrossRef]
3. Matsumoto, Y.; Murakami, M.; Shono, T.; Hasegawa, T.; Fukumura, T.; Kawasaki, M.; Ahmet, P.; Chikyow, T.; Koshihara, S.; Koinuma, H. Room-temperature ferromagnetism in transparent transition metal-doped titanium dioxide. *Science* **2001**, *291*, 854–856. [CrossRef] [PubMed]
4. Serpone, N. Is the band gap of pristine TiO_2 narrowed by anion- and cation-doping of titanium dioxide in second-generation photocatalysts? *J. Phys. Chem. B* **2006**, *110*, 24287–24293. [CrossRef] [PubMed]
5. Vitiello, G.; Pezzella, A.; Calcagno, V.; Silvestri, B.; Raiola, L.; D'Errico, G.; Costantini, A.; Branda, F.; Luciani, G. 5,6-Dihydroxyindole-2-carboxylic acid–TiO_2 charge transfer complexes in the radical polymerization of melanogenic precursor(s). *J. Phys. Chem. C* **2016**, *120*, 6262–6268. [CrossRef]
6. Chiodi, M.; Cheney, C.P.; Vilmercati, P.; Cavaliere, E.; Mannella, N.; Weitering, H.H.; Gavioli, L. Enhanced dopant solubility and visible-light absorption in Cr–N codoped TiO_2 nanoclusters. *J. Phys. Chem. C* **2012**, *116*, 311–318. [CrossRef]
7. Asahi, R.; Morikawa, T.; Ohwaki, T.; Aoki, K.; Taga, Y. Visible-light photocatalysis in nitrogen-doped titanium oxides. *Science* **2001**, *293*, 269–271. [CrossRef]
8. Batzill, M.; Morales, E.H.; Diebold, U. Influence of nitrogen doping on the defect formation and surface properties of TiO_2 rutile and anatase. *Phys. Rev. Lett.* **2006**, *96*, 026103. [CrossRef]
9. Clarizia, L.; Vitiello, G.; Pallotti, D.K.; Silvestri, B.; Nadagouda, M.; Lettieri, S.; Luciani, G.; Andreozzi, R.; Maddalena, P.; Marotta, R. Effect of surface properties of copper-modified commercial titanium dioxide photocatalysts on hydrogen production through photoreforming of alcohols. *Int. J. Hydrogen Energy* **2017**, *42*, 28349–28362. [CrossRef]
10. Zhu, J.; Deng, Z.; Chen, F.; Zhang, J.; Chen, H.; Anpo, M.; Huang, J.; Zhang, L. Hydrothermal doping method for preparation of Cr^{3+}–TiO_2 photocatalysts with concentration gradient distribution of Cr^{3+}. *Appl. Catal. B: Environ.* **2006**, *62*, 329–335. [CrossRef]
11. Fan, X.; Chen, X.; Zhu, S.; Li, Z.; Yu, T.; Ye, J.; Zou, Z. The structural, physical and photocatalytic properties of the mesoporous Cr-doped TiO_2. *J. Mol. Catal. A Chem.* **2008**, *284*, 155–160. [CrossRef]
12. Kaspar, T.C.; Heald, S.M.; Wang, C.M.; Bryan, J.D.; Droubay, T.; Shutthanandan, V.; Thevuthasan, S.; McCready, D.E.; Kellock, A.J.; Gamelin, D.R.; et al. Negligible magnetism in excellent structural quality $Cr_xTi_{1-x}O_2$ anatase: contrast with high-T_C ferromagnetism in structurally defective $Cr_xTi_{1-x}O_2$. *Phys. Rev. Lett.* **2005**, *95*, 217203. [CrossRef] [PubMed]
13. Zhu, W.; Qiu, X.; Iancu, V.; Chen, X.Q.; Pan, H.; Wang, W.; Dimitrijevic, M.N.; Rajh, T.; Meyer, H.M.; Paranthaman, M.P.; et al. Band gap narrowing of titanium oxide semiconductors by noncompensated anion-cation codoping for enhanced visible-light photoactivity. *Phys. Rev. Lett.* **2009**, *103*, 226401. [CrossRef] [PubMed]
14. Wang, Y.; Cheng, Z.; Tan, S.; Shao, X.; Wang, B.; Hou, J.G. Characterization of Cr–N codoped anatase $TiO_2(001)$ thin films epitaxially grown on $SrTiO_3(001)$ substrate. *Surf. Sci.* **2013**, *616*, 93. [CrossRef]
15. Cheng, Z.; Zhang, L.; Dong, S.; Ma, X.; Ju, H.; Zhu, J.; Cui, X.; Zhao, J.; Wang, B. Electronic properties of Cr–N codoped rutile $TiO_2(110)$ thin films. *Surf. Sci.* **2017**, *666*, 84–89. [CrossRef]
16. Kollbek, K.; Szkudlarek, A.; Marzec, M.M.; Lyson-Sypien, B.; Cecot, M.; Bernasik, A.; Radecka, M.; Zakrzewska, K. Optical and electrical properties of $Ti(Cr)O_2$:N thin films deposited by magnetron co-sputtering. *Appl. Surf. Sci.* **2016**, *380*, 73–82. [CrossRef]
17. Gago, R.; Prucnal, S.; Pérez-Casero, R.; Caretti, I.; Jiménez, I.; Lungwitz, F.; Cornelius, S. Structural impact of chromium incorporation in as-grown and flash-lamp-annealed sputter deposited titanium oxide films. *J. Alloy Compd.* **2017**, *729*, 438–445. [CrossRef]
18. Xu, M.; Gao, Y.; Martinez-Moreno, E.; Kunst, M.; Muhler, M.; Wang, Y.; Idriss, H.; Wöll, C. Photocatalytic activity of bulk TiO_2 anatase and rutile single crystals using infrared absorption spectroscopy. *Phys. Rev. Lett.* **2011**, *106*, 138302. [CrossRef]
19. Houska, J.; Mraz, S.; Schneider, J.M. Experimental and molecular dynamics study of the growth of crystalline TiO_2. *J. Appl. Phys.* **2012**, *112*, 073527. [CrossRef]

20. Gago, R.; Redondo-Cubero, A.; Vinnichenko, M.; Lehmann, J.; Munnik, F.; Palomares, F.J. Spectroscopic evidence of NO$_x$ formation and band-gap narrowing in N-doped TiO$_2$ films grown by pulsed magnetron sputtering. *Mater. Chem. Phys.* **2012**, *136*, 729–736. [CrossRef]
21. Skorupa, W.; Gebel, T.; Yankov, R.A.; Paul, S.; Lerch, W.; Downey, D.F.; Arevalo, E.A. Advanced thermal processing of ultrashallow implanted junctions using flash lamp annealing. *J. Electrochem. Soc.* **2005**, *152*, G436–G440. [CrossRef]
22. Mayer, M. *SIMNRA User's Guide 6.05*; Max-Planck-Institut für Plasmaphysik: Garching, Germany, 2009.
23. Barradas, N.P.; Jeynes, C.; Webb, R.P. Simulated annealing analysis of Rutherford backscattering data. *Appl. Phys. Lett.* **1997**, *71*, 291–293. [CrossRef]
24. Stöhr, J. *NEXAFS Spectroscopy*; Springer: New York, NY, USA, 1992.
25. Crocombette, J.P.; Jollet, F. Ti 2p X-ray absorption in titanium dioxides (TiO$_2$): The influence of the cation site environment. *J. Phys. Condens. Matter* **1994**, *6*, 10811. [CrossRef]
26. Gago, R.; Vinnichenko, M.; Redondo-Cubero, A.; Czigány, Z.; Vázquez, L. Surface morphology of heterogeneous nanocrystalline rutile/amorphous anatase TiO$_2$ films grown by reactive pulsed magnetron sputtering. *Plasma Process. Polym.* **2010**, *7*, 813–823. [CrossRef]
27. Lindgren, T.; Mwabora, J.M.; Avendano, E.; Jonsson, J.; Hoel, A.; Granqvist, C.G.; Lindquist, S. Photoelectrochemical and optical properties of nitrogen doped titanium dioxide films prepared by reactive DC magnetron sputtering. *J. Phys. Chem. B* **2003**, *107*, 5709–5716. [CrossRef]
28. Madhavi, V.; Kondaiah, P.; Mohan Rao, G. Influence of silver nanoparticles on titanium oxide and nitrogen doped titanium oxide thin films for sun light photocatalysis. *Appl. Surf. Sci.* **2018**, *436*, 708–719.
29. Mwabora, J.M.; Lindgren, T.; Avendaño, E.; Jaramillo, T.F.; Lu, J.; Lindquist, S.E.; Granqvist, C.G. Structure, composition, and morphology of photoelectrochemically active TiO$_{2-x}$N$_x$ thin films deposited by reactive DC magnetron sputtering. *J. Phys. Chem. B* **2004**, *108*, 20193–20198. [CrossRef]
30. Klug, H.P.; Alexander, L.E. *X-ray Diffraction Procedures for Polycrystalline and Amorphous Materials*; Wiley: Hoboken, NJ, USA, 1974.
31. Ruus, R.; Kikas, A.; Saar, A.; Ausmees, A.; Nommiste, E.; Aarik, J.; Aidla, A.; Uustare, T.; Martinson, I. Ti 2p and O 1s X-ray absorption of TiO$_2$ polymorphs. *Solid State Commun.* **1997**, *104*, 199–203. [CrossRef]
32. Kucheyev, S.O.; Sadigh, B.; Baumann, T.F.; Wang, Y.M.; Felter, T.E.; Van Buuren, T.; Gash, A.E.; Satcher, J.H., Jr.; Hamza, A.V. Electronic structure of chromia aerogels from soft X-ray absorption spectroscopy. *J. Appl. Phys.* **2007**, *101*, 124315. [CrossRef]
33. Schedel-Niedrig, T.; Neisius, T.; Simmons, C.T.; Köhler, K. X-ray absorption spectroscopy of small chromium oxide particles (Cr$_2$O$_3$, CrO$_2$) supported on titanium dioxide. *Langmuir* **1996**, *12*, 6377–6381.
34. Dedkov, Y.S.; Vinogradov, A.S.; Fonin, M.; König, C.; Vyalikh, D.V.; Preobrajenski, A.B.; Krasnikov, S.A.; Kleimenov, E.Y.; Nesterov, M.A.; Rüdiger, U.; et al. Correlations in the electronic structure of half-metallic ferromagnetic CrO$_2$ films: An X-ray absorption and resonant photoemission spectroscopy study. *Phys. Rev. B* **2005**, *72*, 060401(R). [CrossRef]
35. Gago, R.; Vinnichenko, M.; Hübner, R.; Redondo-Cubero, A. Bonding structure and morphology of chromium oxide films grown by pulsed-DC reactive magnetron sputter deposition. *J. Alloy Compd.* **2016**, *672*, 529–535. [CrossRef]

© 2019 by the authors. Licensee MDPI, Basel, Switzerland. This article is an open access article distributed under the terms and conditions of the Creative Commons Attribution (CC BY) license (http://creativecommons.org/licenses/by/4.0/).

Article

The Effect of RF Sputtering Conditions on the Physical Characteristics of Deposited GeGaN Thin Film

Cao Phuong Thao [1], Dong-Hau Kuo [2,*], Thi Tran Anh Tuan [3,*], Kim Anh Tuan [1], Nguyen Hoang Vu [1], Thach Thi Via Sa Na [1], Khau Van Nhut [1] and Nguyen Van Sau [3]

1. School of Engineering and Technology, Tra Vinh University, Tra Vinh 87000, Vietnam; cpthao@tvn.edu.cn (C.P.T.); katuan@tvu.edu.cn (K.A.T.); nghvu@tvu.edu.vn (N.H.V.); viasana@tvu.edu.cn (T.T.V.S.N.); nhutkhau@tvu.edu.vn (K.V.N.)
2. Department of Materials Science and Engineering, National Taiwan University of Science and Technology, Taipei 10607, Taiwan
3. School of Basic Sciences, Tra Vinh University, Tra Vinh 87000, Vietnam; nvsau@tvu.edu.vn
* Correspondence: dhkuo@mail.ntust.edu.tw (D.-H.K.); thitrananhtuan@tvu.edu.vn (T.T.A.T.); Tel.: +886-2-27303291 (D.-H.K.)

Received: 5 September 2019; Accepted: 1 October 2019; Published: 6 October 2019

Abstract: $Ge_{0.07}$GaN films were successfully made on Si (100), SiO_2/Si (100) substrates by a radio frequency reactive sputtering technique at various deposition conditions listed as a range of 100–400 °C and 90–150 W with a single ceramic target containing 7 at % dopant Ge. The results showed that different RF sputtering power and heating temperature conditions affected the structural, electrical and optical properties of the sputtered $Ge_{0.07}$GaN films. The as-deposited $Ge_{0.07}$GaN films had an structural polycrystalline. The GeGaN films had a distorted structure under different growth conditions. The deposited-150 W $Ge_{0.07}$GaN film exhibited the lowest photoenergy of 2.96 eV, the highest electron concentration of 5.50×10^{19} cm^{-3}, a carrier conductivity of 35.2 S·cm^{-1} and mobility of 4 cm^2·V^{-1}·s^{-1}.

Keywords: Ge donor; GaN; growth condition; heating substrate temperature; RF power; reactive sputtering; thin film property

---

1. Introduction

It is known that Gallium Nitride (GaN) and its compounds have wide bandgap, high thermal conductivity [1] and wurtzite crystal structure. They have been employed for electronics and photo-electronic components, listed as MOSFET and HJ-FET transistors, diodes and light emitting diodes (LED) [2–6].

By using the doping technique to make n-type semiconductor materials, Shuji et al. studied the efficiency between Ge and Si doping. The doping of Si had higher efficiency as the GeH_4 and SiH_4 precursors were applied for Ge- and Si-doped GaN with high electron concentrations at 2×10^{19} and 1×10^{19} cm^{-3}, respectively [7]. Ge performed as a charge carrier in GaN film made by a plasma-assisted molecular beam epitaxy (MBE) system [8,9]. Many researchers have applied various deposition techniques to make Ge-doped GaN, such as hydride-vapor phase-epitaxy (HVPE) [10], chemical-vapor-deposition (MOCVD) [7,11,12], metalorganic –vapor- phase-epitaxy (MOVPE) [13,14], and a thermionic-vacuum arc [15].

To investigate the influences of dopant on the semiconductor behaviors, in our previous experiment, we reported Ge-doped GaN film deposited by a radio frequency (RF) reactive sputtering technique with a single ceramic target at the different Ge contents of the dopant of 0, 0.03, 0.07 and 1. It was presented that all these Ge-doped GaN thin films acted in as an n-type semiconductor for the various Ge dopant ratios [16]. Besides, there were many previous works that studied the effects of different sputtering conditions on the doping GaN films [5,17,18]. However, there is not much research exploring the influence of the different growth conditions on Ge-doped GaN film made by the RF reactive sputtering technique until this work. In this research, we study the effects of RF sputtering conditions on properties of these $Ge_{0.07}GaN$ films. Firstly, $Ge_{0.07}GaN$ films were grown at different heating substrate temperatures from 100 to 400 °C. Secondly, the RF sputtering power changed in the range of 90–150 W and was applied to prepare $Ge_{0.07}GaN$ films, while the deposition temperature was fixed at 300 °C.

2. Experimental Details

$Ge_{0.07}GaN$ thin films were successfully deposited on Si (100) substrate by radio-frequency (RF) reactive sputtering with a $Ge_{0.07}GaN$ single ceramic target containing 7 at the % of the Ge/(Ge+Ga) molar ratio. To investigate the influences of deposition temperature, the substrates were heated in a range of 100–400 °C while the output RF power and sputtering time were kept at 120 W and 30 min, respectively. To study the effects of different sputtering powers on properties of $Ge_{0.07}GaN$ films, the films were deposited under 90, 120, and 150 W while the deposition temperature and duration of sputtering were held at 300 °C and 30 min, respectively. The sputtering proceeded under the working pressure at 9×10^{-3} torrs and the mixing gases of Argon flow rate at 5 sccm and Nitrogen flow rate at 15 sccm. The size of the single cermet targets employed in RF sputtering was 5.08 cm (2 inches). The distance between the target and substrates in the working chamber for depositing was kept at 5 cm, while the substrate faced the target. Details for preparing a single ceramic target and RF reactive sputtering process were presented in the previous experiment in our laboratory [16,18–21].

The structural crystallite of the sputtered $Ge_{0.07}GaN$ films deposited under the different heating substrates (range of 100–400 °C) was tested by X-ray diffractometry (XRD, D8 Discover, Bruker, Billerica, MA, USA). The morphological and topographical surfaces of these $Ge_{0.07}GaN$ films were investigated by scanning electron microscopy (SEM, JSM-6500F, JEOL, Tokyo, Japan) and atomic force microscopy (AFM, Dimension Icon, Bruker). The energy dispersive spectrometer (EDS, JSM-6500F, JEOL) prepared on SEM was employed to analyze the composition data of these films. A Hall measurement system (HMS–2000, Ecopia, Tokyo, Japan) including a maximum magnetic-field of 0.51T was applied for electrical properties. An Ultraviolet-Visible (UV-Vis) spectrometer (V-670, Jasco, Tokyo, Japan) was used to study the optical properties of $Ge_{0.07}GaN$ films.

3. Results and Discussion

3.1. Effects of Growth Temperature on the Sputtered GeGaN Film Properties

Compositional EDS investigation of the $Ge_{0.07}GaN$ films deposited in a temperature range from 100 to 400 °C is shown in Table 1. It is shown that the grown $Ge_{0.07}GaN$ films contained nitrogen from 48.4–49.7 at.%, and the [N]/([Ga]+[Ge]) molar ratios were between 0.93–0.98. It was illustrated that these $Ge_{0.07}GaN$ films were composed of slightly deficient nitrogen contents, and inadequate nitrogen was associated with the electrical properties of films. From EDS data displayed in Table 1, [Ge]/([Ge]+[Ga]) molar ratios were 0.057, 0.074, 0.085, and 0.094 for Ge-0.07-GaN films at heating substrate temperatures of 100, 200, 300 and 400 °C, respectively. As the heating substrate temperature increased, there was an increase in the Ge molar ratios of the sputtered $Ge_{0.07}GaN$ films. It was indicated that deposition temperature changed the Ge atom ratio in the deposited film to prove the effect of sputtering temperature on the film properties.

The morphological and topographical surface images of $Ge_{0.07}GaN$ films deposited at different deposition temperatures in the range from 100 to 400 °C are displayed in Figure 1. The SEM surface images indicated that the grown $Ge_{0.07}GaN$ films had a microstructure with continuous and smooth surfaces. From the cross-sectional SEM patterns in Figure 1, these $Ge_{0.07}GaN$ films had a 1.0–1.78 μm thickness and adhered well between Ge$_{0.07}$GaN films and Si wafer with free cracks or voids at interfaces. From data seen in Table 2, as the growth temperature rose from 100 to 400 °C, the growth rate corresponded to 33.33, 39.0, 43.33, and 59.33 nm/min. The root-mean-square (rms) roughness values of these deposited Ge$_{0.07}$GaN films were 1.35, 1.40, 3.0, 3.1 nm as the substrate temperatures in the sputtering process increased from 100 to 400 °C. The sputtered GaN film made by RF sputtering technology had a roughness from 0.7 to 20 nm [22], while the roughness of the GaN films made by the MOCVD method was in the range of 0.5–3 nm [23]. As the deposition temperature changed from 100 to 400 °C, Ge$_{0.07}$GaN films deposited had an increase in roughness value from 1.35 to 3.1 nm and a smooth surface. The morphology of the Ge$_{0.07}$GaN film became rougher as the heating temperature substrate increased. It could be determined that strong bombardment of argon against the Ge$_{0.07}$GaN target at a higher RF sputtering temperature was responsible for the faster deposition rate and the higher roughness of the surface.

Table 1. EDS composition of Ge$_{0.07}$GaN films under different deposition conditions.

Sputtering Conditions		Ga (at.%)	Ge (at.%)	N (at.%)	[Ge]/([Ga]+[Ge]	[N]/([Ga]+[Ge])
Heating Substrate (°C)	100	47.42	2.87	49.71	0.057	0.988
	200	47.08	3.78	49.14	0.074	0.966
	300	46.83	4.35	48.82	0.085	0.954
	400	46.73	4.87	48.40	0.094	0.938
RF Sputtering Power (W)	90	49.39	3.89	46.72	0.073	0.877
	120	46.83	4.35	48.82	0.085	0.954
	150	47.16	5.19	47.65	0.099	0.910

Table 2. The influence of RF power and substrate temperature conditions on the structural properties.

Sputtering Conditions		Film Thickness (μm)	Deposition Rate (nm/minute)	Roughness (nm)
Deposition Temperature (°C)	100	1.00	33.33	1.35
	200	1.17	39.00	1.40
	300	1.30	43.33	3.00
	400	1.78	59.33	3.10
Sputtering Power (W)	90	0.62	20.67	0.46
	120	1.30	43.33	3.00
	150	2.50	83.33	3.77

Figure 2a presented the XRD pattern and slow scan rate spectra of the Ge$_{0.07}$GaN films grown by RF sputtering at a different heating temperature in the range of 100–400 °C and at 120 W of RF power under the mixing of Ar/N$_2$ input gases. From the surveyed XRD, all Ge$_{0.07}$GaN films deposited on Si (100) substrates at temperature 100–400 °C were polycrystalline including structural wurtzite, and Ge constituted the solid-state solution in the GaN crystal structure [16]. It could be clearly seen that these Ge$_{0.07}$GaN films with a preferential ($10\bar{1}0$) growth plane had ($10\bar{1}0$), ($10\bar{1}1$), ($11\bar{2}0$) and ($11\bar{2}2$) peaks, and other secondary phases could not be found. At the higher heating temperature, the ($10\bar{1}0$) peak slightly shifted to the higher 2θ angle, and the ($10\bar{1}0$) peak of the deposited Ge$_{0.07}$GaN films at 100, 200, 300 and 400 °C was located at 32.25°, 32.30°, 32.36° and 32.40°, respectively. Table 2 shows the parameters for the crystal structure of Ge$_{0.07}$GaN films grown at different temperatures. The lattice constant c slightly decreased from 5.21, 5.20, 5.18 to 5.17 Å and a was 3.21, 3.20, 3.19 and 3.18 Å, corresponding to GeGaN films made at the heating substrate temperatures of 100, 200, 300 and 400 °C, respectively. Additionally, cell volumes of Ge-0.07-GaN films sputtered at 100, 200, 300, and 400 °C were 46.57, 46.13, 45.70 and 45.27 at Å^3, respectively. From XRD data in Table 3, the dominant ($10\bar{1}0$) peaks of the 100–400 °C deposited Ge$_{0.07}$GaN films were slightly reduced with respect to the full width at half maxima (FWHM) values, i.e., 0.34°, 0.30°, 0.27° and 0.25° at 100, 200, 300, and 400 °C, respectively. Additionally, the crystalline size could be computed by the Scherer equation and the was significantly greater at higher heating temperature: 24.33, 27.57, 30.64, and 33.09 nm for the Ge$_{0.07}$GaN films deposited at 100, 200, 300, and 400 °C, respectively. It could be believed that the heating temperature affected the structural crystallite of the film as the Ge$_{0.07}$GaN films were deposited by RF sputtering at 100, 200, 300, and 400 °C.

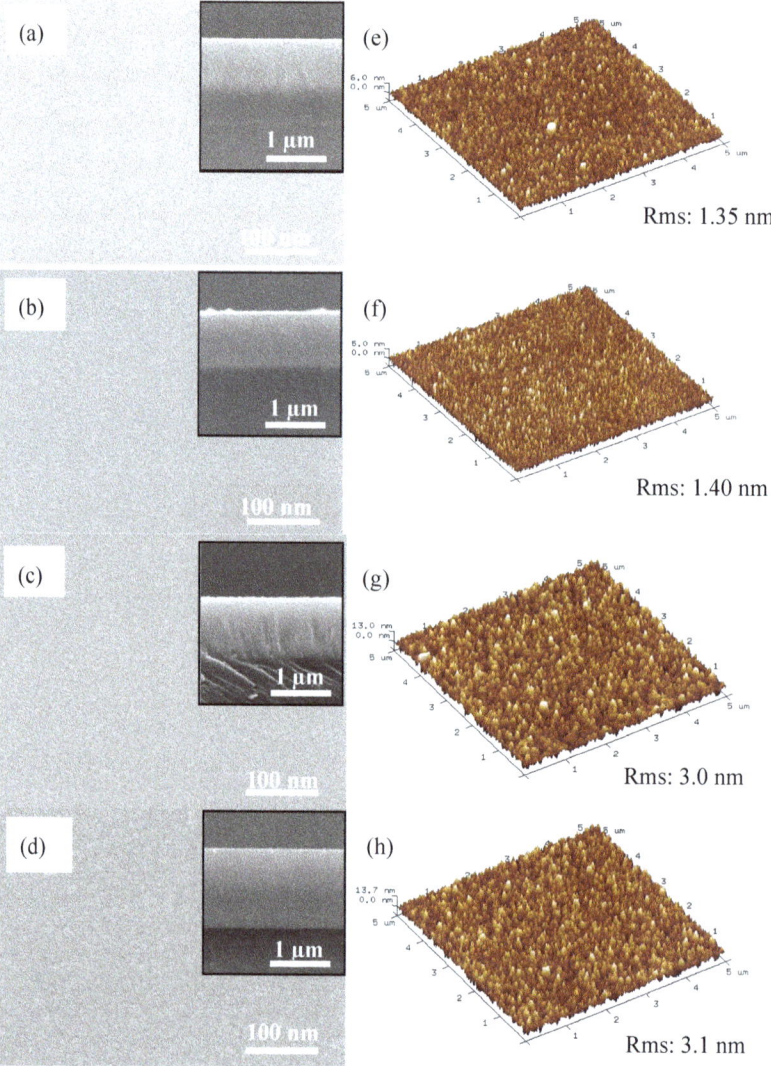

Figure 1. (**a–d**) SEM surface images and (**e–h**) 3D AFM morphologies of Ge$_{0.07}$GaN films at (**a,e**) 100 °C, (**b,f**) 200 °C, (**c,g**) 30 and 400 °C. The insets are their individual cross–sectional images.

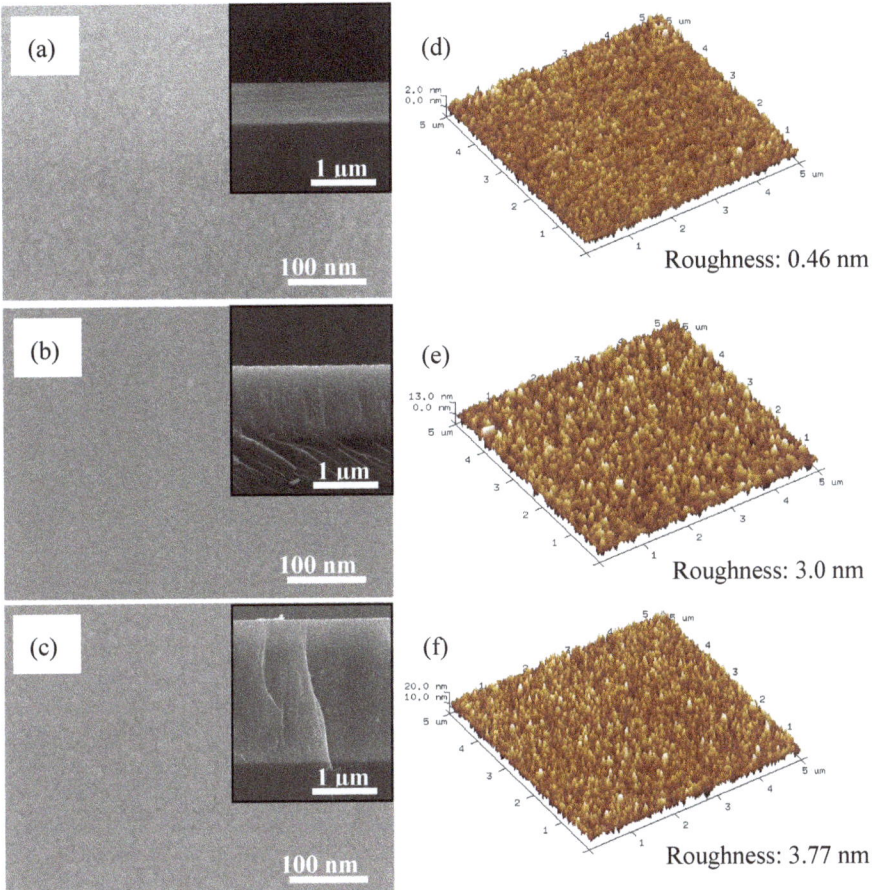

Figure 2. (a,b,d) SEM surface images and (e–g) 3D AFM morphologies of Ge$_{0.07}$GaN films deposited at (a,d) 90 W, (b,e) 120 W, and (c,f) 150 W in Ar/N$_2$ atmosphere. The insets present their single cross-sectional images.

Table 3. Structure properties of Ge$_{0.07}$GaN thin films at different sputtering powers and substrate temperature from X-ray diffraction analyses.

Sputtering Conditions		2θ(1010) peak	a (Å)	c (Å)	Volume (Å^3)	FWHM (1010) (degree)	Crystallite Size (nm)
Deposition Temperature (°C)	100	32.25	3.21	5.21	46.57	0.34	24.33
	200	32.30	3.20	5.20	46.13	0.30	27.57
	300	32.36	3.19	5.18	45.70	0.27	30.64
	400	32.40	3.18	5.17	45.27	0.25	33.09
RF Sputtering Power (W)	90	32.30	3.59	5.83	64.87	0.41	21.44
	120	32.36	3.38	5.49	54.26	0.27	33.04
	150	32.40	3.25	5.29	48.43	0.26	34.67

The electrical properties of Ge$_{0.07}$GaN films deposited at different temperatures in the range of 100–400 °C and the 120 W of RF power were investigated by the Hall measurement system. In previous experiments, we reported that the 300 °C-sputtered Ge$_{0.07}$GaN film achieved an electron concentration of 5.02×10^{17} cm^{-3}, mobility of 10.5 cm$^2 \cdot$V$^{-1} \cdot$s^{-1}, and carrier conductivity of 10.84 S·cm^{-1}, and worked as an n-semiconductor layer [16]. From data displayed in Table 4 and Figure 3a, all sputtered $Ge_{0.07}GaN$ films at different growth temperatures from 100 to 400 °C remained n-type semiconductors. It could be explained that the compositional EDS data shown in Table 1 were responsible for the electrical properties of the $Ge_{0.07}GaN$ films. The $Ge_{0.07}GaN$ film at 100, 200, 300, and 400 °C had an increase in electron concentration (n_e) from 1.64×10^{16}, 2.14×10^{17}, 5.02×10^{17} to 1.30×10^{18} cm^{-3}, and a decrease in mobility (μ) between 33, 17, 11 and 7 cm$^2 \cdot V^{-1} \cdot s^{-1}$, respectively. It is believed that electron concentration could be a function of electrical conductivity and the as-deposited GeGaN at 100, 200, 300 and 400 °C maintained the increase in electronic conductivity corresponding to 0.09, 0.58, 0.88 and 1.46 S·cm^{-1}, respectively. The practical electrical properties of these $Ge_{0.07}GaN$ films illustrated that there were effects of heating substrate temperatures on film properties.

Table 4. Electrical properties of $Ge_{0.07}GaN$ films deposited at different temperatures.

Sputtering Conditions.		Type	Concentration N_e cm^{-3}	Mobility μ cm$^2 \cdot V^{-1} \cdot s^{-1}$	Conductivity σ S·cm^{-1}	Bandgap eV
Deposition Temperature (°C)	100	n	1.64×10^{16}	33	0.09	3.14
	200	n	2.14×10^{17}	17	0.58	3.09
	300	n	5.02×10^{17}	11	0.88	3.05
	400	n	1.30×10^{18}	7	1.46	3.02
Sputtering Power (W)	90	n	3.22×10^{15}	25	0.012	3.14
	120	n	5.02×10^{17}	11	0.84	3.05
	150	n	5.50×10^{19}	4	35.2	2.96

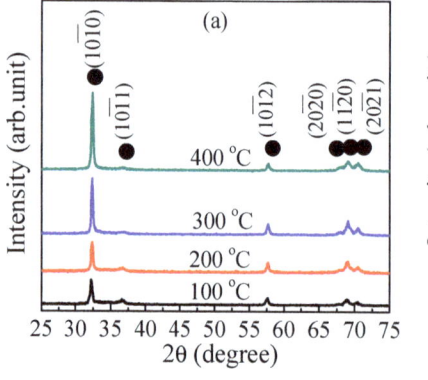
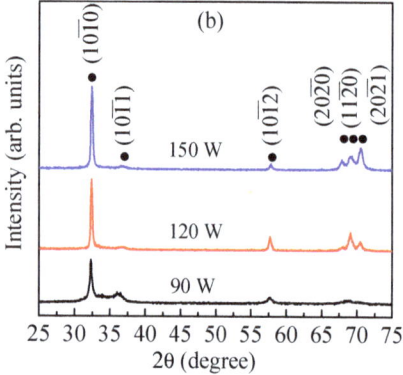

Figure 3. XRD patterns of $Ge_{0.07}GaN$ films deposited at (a) different growth temperatures; (b) different RF power (90–150 W) in an Ar/N$_2$ atmosphere.

The absorption of GeGaN films was studied by UV–Vis measurement at room temperature. The Equation (1) named the Tauc equation has been used to show the optical absorption coefficient and energy bandgap (E_g) of $Ge_{0.07}GaN$ films from the UV–Vis database.

$$(\alpha h\nu)^2 = A (h\nu - E_g) \quad (1)$$

where A is a invariable number, α is the coefficient of optical absorption. From equation, the incident photon and the $Ge_{0.07}GaN$ films bandgap of energies were determined and listed for $h\nu$ and E_g. Figure 4a and Table 4 show the plots of the $(\alpha h\nu)^2 - h\nu$ curves and the bandgap values of $Ge_{0.07}GaN$ films deposited at different temperatures, which could be directly obtained by extrapolating the linear part of these curves. The E_g values from the extrapolated curves were 3.14, 3.09, 3.05, and 3.02 eV for $Ge_{0.07}GaN$ films deposited at different temperatures from 100 to 400 °C.

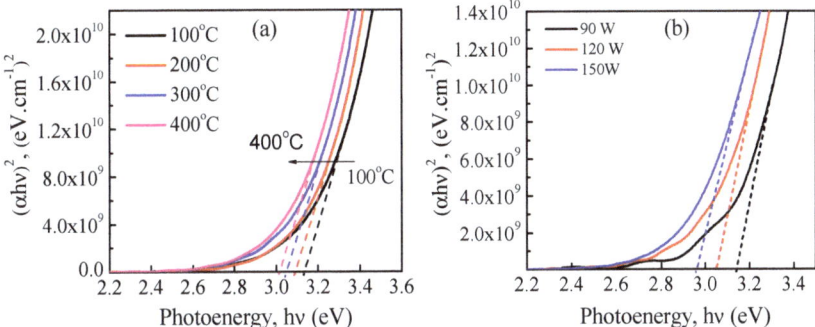

Figure 4. Plots of $(\alpha h\nu)^2$ vs. photon energy ($h\nu$) for the optical band gap determination of the Ge$_{0.07}$GaN films sputtered in a (**a**) deposition temperature (100–400 °C) range; (**b**) RF reactive sputtering range of 90–150 W.

3.2. Influences of RF Sputtering Power on the Electrical, Optical and Structural Properties of Ge-Doped Gan Thin Films

The composition of Ge$_{0.07}$GaN films as-deposited at 90, 120 and 150 W RF sputtering power is shown in Table 1. The ratios of molar [Ge]/([Ge]+[Ga]) were 7.3, 8.5 and 9.9 at % for Ge$_{0.07}$GaN films grown at 90, 120 and 150 W, respectively. Under output RF power conditions, the Ge content in sputtered films increased with the RF power. Moreover, the nitrogen contents in these films were less than 50 at %, which indicates that there was a nitrogen-deficiency state in Ge$_{0.07}$GaN films at different sputtering powers.

The surface morphology and cross-section images of Ge-GaN films grown at different output RF sputtering powers are presented in Figure 4. The results of SEM images showed the smoothness surface and grains in nanometer size without voids and mechanical fracture phenomena. It is found that the higher sputtering power of deposition processes is the reason for the crystal grains having sufficient energy, causing the increase in the size of grains. From the cross-sectional patterns of Ge$_{0.07}$GaN films at 90–150 W of RF power in Figure 5, film thickness increased from 0.62 to 2.5 μm and explained the excellent adhesion, and no cracks or holes appeared at the surface between Si substrate and films. It can be observed that the thickness of the film increased as the film was deposited under a higher sputtering power. This means that the sputtering rate increased because the number of atoms deposited on the substrate increase and the film thickness will become thicker. From data in Table 3, the sputtering growth rate was 20.67 43.33 and 83.33 nm/min corresponding to 90, 120 and 150 W of deposition power. This experiment successful prepared Ge$_{0.07}$GaN films under different sputtering powers without a buffer layer film. It can be seen from Figure 2 that GeGaN films have increased grain size as the RF sputtering power increased from 90 to 150 W, which is due to the higher the sputtering power, the higher the current density of the plasma, and the free energy of the gas molecules, which increases, so that the opportunity to effectively hit the target increased, while the sputtered atoms have a large kinetic energy, arrived at the substrate with a high surface energy for grain growth, and increased the grain size. Using the Scherer equation, the crystalline size could be 24.33, 27.57, 30.64, and 33.09 nm for the Ge$_{0.07}$GaN films grown at 90, 120, and 150 W, respectively.

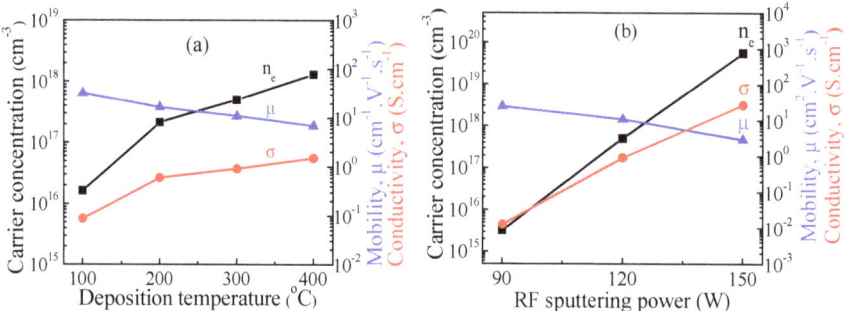

Figure 5. Electrical properties of $Ge_{0.07}GaN$ films deposited under a (**a**) heating substrate temperature range of 100–400 °C, (**b**) RF sputtering power from 90 to 150 W.

Under deposition power conditions of 90, 120 and 150 W, the roughness of Ge-doped GaN films was 0.46, 3.0 and 3.2 nm, respectively. It is explained that there is a relationship between the increase in sputtering power and the surface roughness of the film. The result of the roughness of films showed that higher bombardment of atoms from the target resulted in an increase in the deposition rate at higher sputtering power. Under higher output power, atoms have high surface movement energy to cause coarsening of grains and increases in the surface roughness of films.

Figure 3b shows the XRD images of $Ge_{0.07}GaN$ thin films deposited under different output powers of 90, 120 and 150 W. The XRD results show that these $Ge_{0.07}GaN$ films exhibited a wurtzite structure. At the higher RF power of 120 and 150 W, the sputtered $Ge_{0.07}GaN$ films were polycrystalline. However, there was deficient momentum between atoms and the substrate during the depositing process at 90 W RF power with respect to the low-quality crystallite of the $Ge_{0.07}GaN$ film. The $(10\bar{1}0)$, $(10\bar{1}1)$, $(10\bar{1}2)$, $(11\bar{2}0)$ and $(11\bar{2}2)$ diffraction peaks were found in Ge-GaN films, and no other secondary phase was detected. The peak positions of the $(10\bar{1}0)$ lattice plane were located at 32.30°, 32.36° and 32.40° as the sputtering power was kept at 90, 120, and 150 W, respectively. The 2θ angle of diffraction peaks slightly shifted higher at higher power. Table 2 shows all the calculated data from the XRD investigation. The a, c lattice constants and a unit cell volume of Ge-GaN films slightly degraded at higher RF power. While the c lattice constant slightly dropped from 5.83 Å, 5.49 Å, and to 5.29 Å, there was a reduction in the a lattice constant from 3.59 Å, 3.38 Å, and to 3.25 Å, with the cell volume of 64.87 Å^3, 54.26 Å^3 to 48.43 Å^3 corresponding to 90, 120 and 150 W, respectively. The full-width-half-maximum (FWHM) values of the $(10\bar{1}0)$ diffraction peaks of the 2θ value decreased from 0.41° for 90 W power to 0.26° for 150 W power. From XRD investigation at the higher sputtering power, the $Ge_{0.07}GaN$ films achieved a higher crystallinity quality. The XRD of $Ge_{0.07}GaN$ film deposited at 90 W showed the worst crystallinity because the low sputtering power condition created less Ge in the film. All the evidence indicates the formation of GeGaN films was affected by different sputtering powers.

Electrical properties of $Ge_{0.07}GaN$ films sputtered under output powers of 90, 120, and 150 W were investigated by the Hall effect measurement system at room temperature. The electron concentration (n_e), mobility (μ), and conductivity (σ) are plotted in Figure 4b and shown in Table 4. All the $Ge_{0.07}GaN$ films deposited under different output powers presented as a semiconductor of the n-type. The electrical concentration (n_e) was 3.22×10^{15}, 5.02×10^{17} and 5.50×10^{19} cm^{-3} while the electron mobility (μ) was 25, 11 and 4 cm$^2 \cdot V^{-1} \cdot s^{-1}$ at the sputtering power of 90, 120, and 150 W, respectively. The results from the experiments showed that carrier concentration increased with sputtering power. It can be explained that the sputtering power provides energy to the Ge solid solution in the GaN lattice. At low sputtering power, insufficient Ge solid solution precipitated at the grain boundaries prevents internal carrier transfer in films, which causes a lower free carrier concentration while Ge solid solution can be increased with the power upgrade. Additionally, the electrical conductivity (σ) of the films was affected by carrier concentration (n_e) and mobility (μ), and electrical conductivity (σ) was 0.012, 0.84 and 35.2 S·cm^{-1}. The data show that the electrical conductivity increases as the output power increases.

The absorption coefficient and optical bandgap (E_g) of Ge$_{0.07}$GaN films deposited at room temperature on a transparent glass plate at 90–150 W were tested by UV-Vis spectrometry. Figure 5b shows the extrapolated linear part of the $(\alpha h\nu)^2 - h\nu$ curves from which the optical bandgap of Ge$_{0.07}$GaN films could be directly achieved, and the energy bandgap E_g was 3.14, 3.05, 2.96 eV for Ge$_{0.07}$GaN films under power conditions of 90, 120 and 150 W, respectively. As the sputtering power increased, the energy gap gradually became smaller and decreased by 0.18 eV from 90 watts to 150 watts. It is concluded that the increase in the RF power supplied sufficient energy to dissolve the Ge atoms into the lattice of GaN, resulting in a decrease in the energy gap. As a result of electrical properties, it can be found that the carrier concentration increased with the increase in the sputtering power, and the film deposited at 150 watts has the highest carrier concentration and teh minimum energy gap.

4. Conclusions

Ge$_{0.07}$GaN films were deposited on Si (100) substrates by employing radio frequency reactive magnetron sputtering technology at different temperature and RF power conditions. The characteristics and microstructure of these GeGaN films were studied thoroughly by AFM, SEM, XRD, UV–Vis spectrometry and the Hall effect measurement. The results showed that the Ge$_{0.07}$GaN films remained in the polycrystalline structure and conductivity under the different growth conditions. The various sputtering conditions of the deposition process affected structural GeGaN films and resulted in heavy structural distortion. Compared with the sputtered film at different RF power values in the range of 90–150 W, the sputtered-150 W Ge$_{0.07}$GaN films achieved the lowest energy bandgap of 2.96 eV, the highest carrier concentration of 5.50×10^{19} cm^{-3} and electrical conductivity of 35.2 S·cm^{-1}, and 4 cm^2·V^{-1}·s^{-1} mobility. Besides, the analysis of Ge$_{0.07}$GaN films at different substrate temperatures proved the influences of deposition temperature on the structure and properties of the films. From all investigated data, it could be believed that growth conditions of the RF reactive sputtering process affected the structure and properties of Ge$_{0.07}$GaN films.

Author Contributions: Data curation, C.P.T. and T.T.A.T.; Methodology; Writing—original draft, investigation, C.P.T. and T.T.A.T., Formal analysis, Funding acquisition, Writing—review & editing, C.P.T., T.T.A.T., N.H.V., K.A.T., T.T.V.S.N., K.V.N., and N.V.S.; Supervision, D.-H.K.

Funding: This research was funded by the Ministry of Science and Technology of the Republic of China under grant number 107-2221-E-011-141-MY3.

Conflicts of Interest: The authors declare no conflict of interest.

References

1. Akasaki, I.; Amano, H. Crystal growth and conductivity control of group III nitride semiconductors and their application to short wavelength light emitters. *Jpn. J. Appl. Phys.* **1997**, *36*, 5393–5408. [CrossRef]
2. Fujii, T.; Gao, Y.; Sharma, R.; Hu, E.L.; DenBaars, S.P.; Nakamura, S. Increase in the extraction efficiency of GaN-based light-emitting diodes via surface roughening. *Appl. Phys. Lett.* **2004**, *84*, 855–857. [CrossRef]
3. Pearton, S.J.; Ren, F.; Zhang, A.P.; Lee, K.P. Fabrication and performance of GaN electronic devices. *Mater. Sci. Eng. R Rep.* **2000**, *30*, 205–212. [CrossRef]
4. Tuan, T.T.A.; Kuo, D.-H. Characteristics of RF reactive sputter-deposited Pt/SiO$_2$/n-InGaN MoS Schottky diodes. *Mater. Sci. Semicond. Process.* **2015**, *30*, 314–320. [CrossRef]
5. Tuan, T.T.A.; Kuo, D.-H.; Saragih, A.D.; Li, G.-Z. Electrical properties of RF-sputtered Zn-doped GaN films and p-Zn-GaN/n-Si hetero junction diode with low leakage current of 10^{-9} A and a high rectification ratio above 10^5. *Mater. Sci. Eng. B* **2017**, *222*, 18–25. [CrossRef]
6. Kuo, D.-H.; Liu, Y.-T. Characterization of quaternary Zn/Sn-codoped GaN films obtained with Zn$_x$Sn$_{0.04}$GaN targets at different Zn contents by the RF reactive magnetron sputtering technology. *J. Mater. Sci.* **2018**, *53*, 9099–9106. [CrossRef]

7. Nakamura, S.; Mukai, T.; Senoh, M. Si- and Ge-doped GaN films grown with GaN buffer layers. *Jpn. J. Appl. Phys.* **1992**, *31*, 2883–2888. [CrossRef]
8. Hageman, P.R.; Schaff, W.J.; Janinski, J.; Liliental-Weber, Z. n-type doping of wurtzite GaN with germanium grown with plasma-assisted molecular beam epitaxy. *J. Cryst. Growth.* **2004**, *267*, 123–128. [CrossRef]
9. Colussi, M.L.; Baierle, R.J.; Miwa, R.H. Doping effects of C, Si and Ge in wurtzite [0001] GaN, AlN, and InN nanowires. *J. Appl. Phys.* **2011**, *110*, 033709. [CrossRef]
10. Oshima, Y.; Yoshida, T.; Watanabe, K.; Mishima, T. Properties of Ge-doped, high-quality bulk GaN crystals fabricated by hydride vapor phase epitaxy. *J. Cryst. Growth.* **2010**, *312*, 3569–3573. [CrossRef]
11. Kirste, R.; Hoffmann, M.P.; Sachet, E.; Bobea, M.; Bryan, Z.; Bryan, I.; Nenstiel, C.; Hoffmann, A.; Maria, J.-P.; Collazo, R.; et al. Ge doped GaN with controllable high carrier concentration for plasmonic applications. *Appl. Phys. Lett.* **2013**, *103*, 242107. [CrossRef]
12. Shikanaia, A.; Fukahori, H.; Kawakami, Y.; Hazu, K.; Sota, T.; Mitani, T.; Mukai, T.; Fujita, S. Optical properties of Si-, Ge- and Sn-doped GaN. *Phys. Status. Solidi. B* **2003**, *235*, 26–30. [CrossRef]
13. Dadgar, A.; Bläsing, J.; Diez, A.; Krost, A. Crack-free, highly conducting GaN layers on Si substrates by Ge doping. *Appl. Phys. Express.* **2011**, *4*, 011001. [CrossRef]
14. Fritze, S.; Dadgar, A.; Witte, H.; Bügler, M.; Rohrbeck, A.; Bläsing, J.; Hoffmann, A.; Krost, A. High Si and Ge n-type doping of GaN doping-Limits and impact on stress. *Appl. Phys. Lett.* **2012**, *100*, 122104. [CrossRef]
15. Özen, S.; Korkmaz, Ş.; Şenay, V.; Pat, S. The substrate effect on Ge doped GaN thin films coated by thermionic vacuum arc. *J. Mater. Sci.: Mater. Electron.* **2017**, *28*, 1288–1293. [CrossRef]
16. Thao, C.P.; Kuo, D.H. Electrical and structural characteristics of Ge-doped GaN thin films and its hetero-junction diode made all by RF reactive sputtering. *Mater. Sci. Semicond Process* **2018**, *74*, 336–341. [CrossRef]
17. Li, C.-C.; Kuo, D.-H. Effects of growth temperature on electrical and structural properties of sputtered GaN films with a cermet target. *J. Mater. Sci.: Mater. Electron.* **2014**, *25*, 1404–1409. [CrossRef]
18. Thao, C.P.; Kuo, D.-H.; Jan, D.-J. Codoping effects of the Zn acceptor on the structural characteristics and electrical properties of the Ge donor-doped GaN thin films and its hetero-junction diodes all made by reactive sputtering. *Mater. Sci. Semicond. Process.* **2018**, *82*, 126–134. [CrossRef]
19. Li, C.-C.; Kuo, D.-H. Material and technology developments of the totally sputtering-made p/n GaN diodes for cost-effective power electronics. *J. Mater. Sci.: Mater. Electron.* **2014**, *25*, 1942–1948. [CrossRef]
20. Tuan, T.T.A.; Kuo, D.-H.; Li, C.C.; Yen, W.-C. Schottky barrier characteristics of Pt contacts to all sputtering-made n-type GaN and MoS diodes. *J. Mater. Sci.: Mater. Electron.* **2014**, *25*, 3264–3270. [CrossRef]
21. Ting, C.-W.; Thao, C.P.; Kuo, D.-H. Electrical and structural characteristics of tin-doped GaN thin films and its hetero-junction diode made all by RF reactive sputtering. *Mater. Sci. Semicond Process.* **2017**, *59*, 50–55. [CrossRef]
22. Kim, H.W.; Kim, N.H. Preparation of GaN films on Zno buffer layers by rf magnetron sputtering. *Appl. Surf. Sci.* **2004**, *236*, 192–197. [CrossRef]
23. Chyr, I.; Lee, B.; Chao, L.C.; Steckl, A.J. Damage generation and removal in the Ga$^+$ focused ion beam micromachining of GaN for photonic applications. *J. Vac. Sci. Technol.* **1999**, *17*, 3063–3067. [CrossRef]

© 2019 by the authors. Licensee MDPI, Basel, Switzerland. This article is an open access article distributed under the terms and conditions of the Creative Commons Attribution (CC BY) license (http://creativecommons.org/licenses/by/4.0/).

Article

Electrical and Structural Properties of All-Sputtered Al/SiO₂/*p*-GaN MOS Schottky Diode

Tran Anh Tuan Thi [1,*], Dong-Hau Kuo [2,*], Phuong Thao Cao [3], Pham Quoc-Phong [3], Vinh Khanh Nghi [3] and Nguyen Phuong Lan Tran [4]

1. School of Basic Sciences, Tra Vinh University, Tra Vinh 87000, Vietnam
2. Department of Materials Science and Engineering, National Taiwan University of Science and Technology, Taipei 10607, Taiwan
3. School of Engineering and Technology, Tra Vinh University, Tra Vinh 87000, Vietnam; cpthao@tvu.edu.vn (P.T.C.); phongpham@tvu.edu.vn (P.Q.-P.); nghivinhkhanh@tvu.edu.vn (V.K.N.)
4. School of Engineering and Technology, Can Tho University, Can Tho 94000, Vietnam; tnplan@ctu.edu.vn
* Correspondence: thitrananhtuan@tvu.edu.vn (T.A.T.T.); dhkuo@mail.ntust.edu.tw (D.-H.K.); Tel.: +886-2-2730-3291 (D.-H.K.)

Received: 22 September 2019; Accepted: 14 October 2019; Published: 21 October 2019

Abstract: The all-sputtered Al/SiO$_2$/*p*-GaN metal-oxide-semiconductor (MOS) Schottky diode was fabricated by the cost-effective radio-frequency sputtering technique with a cermet target at 400 °C. Using scanning electron microscope (SEM), the thicknesses of the electrodes, insulator SiO$_2$ layer, and *p*-GaN were found to be ~250 nm, 70 nm, and 1 µm, respectively. By Hall measurement of a *p*-Mg-GaN film on an SiO$_2$/Si (100) substrate at room temperature, the hole's concentration (N_p) and carrier mobility (µ) were found to be $N_p = 4.32 \times 10^{16}$ cm^{-3} and µ = 7.52 cm$^2 \cdot$V$^{-1} \cdot$s^{-1}, respectively. The atomic force microscope (AFM) results showed that the surface topography of the *p*-GaN film had smoother, smaller grains with a root-mean-square (rms) roughness of 3.27 nm. By *I–V* measurements at room temperature (RT), the electrical properties of the diode had a leakage current of ~4.49 × 10^{-8} A at −1 V, a breakdown voltage of −6 V, a turn-on voltage of ~2.1 V, and a Schottky barrier height (SBH) of 0.67 eV. By *C–V* measurement at RT, with a frequency range of 100–1000 KHz, the concentration of the diode's hole increased from 3.92 × 10^{16} cm^{-3} at 100 kHz to 5.36 × 10^{16} cm^{-3} at 1 MHz, while the Fermi level decreased slightly from 0.109 to 0.099 eV. The SBH of the diode at RT in the *C–V* test was higher than in the *I–V* test because of the induced charges by dielectric layer. In addition, the ideality factor (*n*) and series resistance (R_s) determined by Cheung's and Norde's methods, other parameters for MOS diodes were also calculated by *C–V* measurement at different frequencies.

Keywords: MOS Schottky diode; SBH; *I–V* measurement; *C–V* measurement; Cheung's and Norde's methods

1. Introduction

GaN-based semiconductor materials are currently of interest for the fabrication of electronic devices such as the metal-semiconductor (MS) and MOS Schottky diodes, light-emitting diodes (LEDs), photo-detector, metal-oxide-semiconductor field-effect transistors (MOSFETs), and heterojunction field-effect transistors (HFETs). [1–5]. Previous studies created the thin, high-quality insulator layer between the metal and semiconductor that is used to create a metal-oxide-semiconductor (MOS) structure, which was an important factor for the high-performance of MOS devices [6–10]. Researchers investigated the contact of MOS layers via various approaches, e.g., Al/HfO$_2$/*p*-Si [7], Pt/oxide/*n*-InGaP [10], Pt/SiO$_2$/*n*-InGaN [11], Pd/NiO/GaN [12], Au/SiO$_2$/*n*-GaN [13], Au/SnO$_x$/*n*-LTPS/glass [14], Pt/SiO$_2$/*n*-GaN [6,15], Pt/Oxide/Al$_{0.3}$Ga$_{0.7}$As [16],

Pd/HfO$_2$/GaN [17], and Al/SnO$_2$/p-Si (111) [18]. Due to the presence of the oxide layer, several parameters can be applied to improve the characteristics of electronic devices. Bengi et al. reported the parameters of the Al/HfO$_2$/p-Si MOS device, which was tested by C–V measurement. Their SBHs were shown from 0.17 to 0.98 eV, in the temperature range 300–400 K [7]. Karadeniz et al. investigated the Al/SnO$_2$/p-Si (111) diode using spray deposition. The MOS diode showed a Schottky barrier height (SBH) of 0.52 V, an ideality factor of 2.4, and series resistance of 66 Ω [18]. Liu et al. studied the influence of hydrogen adsorption on the Pd/AlGaN-based MOS diode with SiO$_2$ passivation [19]. Their SBHs were reduced from 0.98 to 0.75 eV under exposure to a 1% H$_2$/air gas.

In this study, the radio-frequency (RF) reactive sputtering technique was used to design the Al/SiO$_2$/p-GaN MOS Schottky diode because of advantages such as low deposition temperature, low cost, and safety [3,6,11]. With the support of the parameters and using the RF technique, our diode was fabricated below 400 °C. The characteristics of the MOS Schottky diode were tested using I–V and C–V measurements. The parameters of the diode were calculated by thermionic emission (TE) mode using Cheung's and Norde's methods.

2. Materials and Methods

Figure 1 shows the modeling of the Al/SiO$_2$/p-GaN MOS Schottky diode based on p-GaN film. First, for the Schottky contacts, an Al layer was sputtered on an SiO$_2$/Si (100) substrate at 200 °C for 20 min using a pure Al (99.99%) target, and RF power of 80 W. To construct the MOS Schottky diodes, an interlayer between Al and p-GaN was designed by depositing SiO$_2$. The SiO$_2$ film was sputtered on an Al/SiO$_2$/Si (100) substrate at 100 °C for 10 min using a quartz target. The RF power remained at 80 W and the Ar atmosphere at a flow rate of 5 sccm. Second, the Mg-GaN films were deposited onto SiO$_2$/Al/SiO$_2$/Si (100) and SiO$_2$/Si (100) substrates at 400 °C for 40 min. The RF power of was kept at 150 W with a gas mixture of Ar and N$_2$ and a flow rate of 5 sccm for each. The 2-inch Mg-GaN target had an [Mg]/([Ga] + [Mg]) molar ratio of 10% and was made via hot pressing. Finally, a Pt-Omhic contact with a size of 1 mm^2 was deposited, at 200 °C for 20 min, with a pure Pt (99.99%) target using a stainless mask.

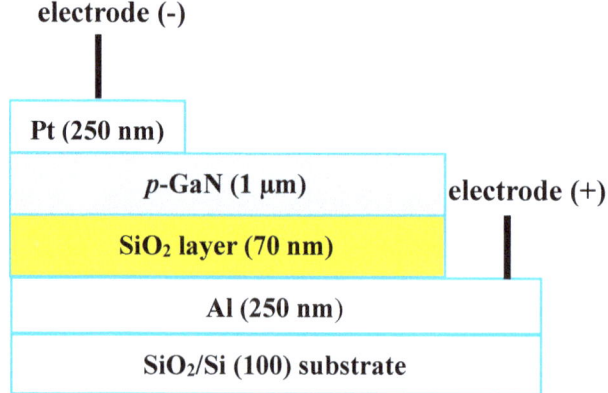

Figure 1. The modeling of the as-deposited Al/SiO$_2$/p-GaN MOS Schottky diode.

The composition analysis and surface topographies of the p-GaN films were determined via SEM and EDS (JSM-6500F, JEOL, Tokyo, Japan), AFM (Dimension Icon, Bruker, Tokyo, Japan). The hole's concentration (N_p) and the mobilities (μ) of the p-GaN film were calculated by Hall measurement (HMS-2000, Ecopia, Tokyo, Japan). The I–V and C–V measurement of the MOS Schottky diode were tested using a semiconductor device analyzer (Agilent, B1500A, Santa Clara, CA, USA) at RT. All the parameters of the MOS Schottky diode were considered by thermionic emission (TE) mode using Cheung's and Norde's methods.

3. Results and Discussion

3.1. Structural and Surface Morphological Characteristics

By Hall measurement of the p-Mg-GaN film on the SiO_2/Si (100) substrate at RT, the hole's concentration (N_p) and carrier mobility (μ) were found to be N_p = 4.32 × 10^{16} cm^{-3} and μ = 7.52 cm$^2 \cdot$V$^{-1} \cdot$s^{-1}, respectively. Using SEM, the thicknesses of both the electrodes and the SiO_2 layer were found to be 250 and 70 nm, respectively.

Figure 2a shows the SEM surface morphologies of the p-GaN films sputtered on the SiO_2/Si (100) substrate. With EDS analysis results, the ratio of [Mg]/([Ga] + [Mg]) was 10.2% for the p-GaN film. This indicated that the p-Mg-GaN film deposited at 400 °C with up to 10% Mg displayed continuous smoothness without cracks and pores. The inset shows a cross-sectional image, with a thickness of 1 μm for the p-GaN film. Figure 2b shows the surface topography of the as-deposited Mg-GaN films on the SiO_2/Si (100) substrate tested by AFM measurement. The surface topography showed smoother and smaller grains and the root-mean-square (rms) roughness of the films was found to be 3.27 nm. The EDS compositions, SEM surface morphologies and XRD patterns of the p-Mg-GaN film obtained with cermet targets at different Mg contents can be found in our previous works [20,21]. The positive surface conditions of Mg-GaN layer together with the insulator SiO_2 layer were the important factors for determining the electrical properties of the MOS Schottky diodes.

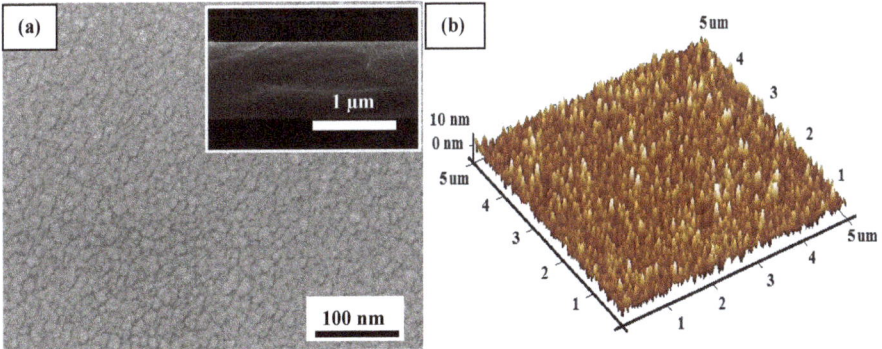

Figure 2. (a) SEM surface image and (b) three-dimensional AFM topographies of the p-GaN film deposited on the SiO_2/Si (100) substrate. The inset is the cross-sectional image of the p-GaN film.

3.2. Current–Voltage (I–V) Characteristics

Figure 3a displays the I–V plot of the Al/SiO_2/p-GaN MOS Schottky diode measured at RT. The Figure 3b shows the ln$I$–V semilogarithmic view of the diode. From the I–V data, tested with a voltage range of (−6 V; +6 V) and a leakage current of −1 V, the turn-on voltage of the diode was determined to be ~4.49 × 10^{-6} A/cm^2 and 2.3 V.

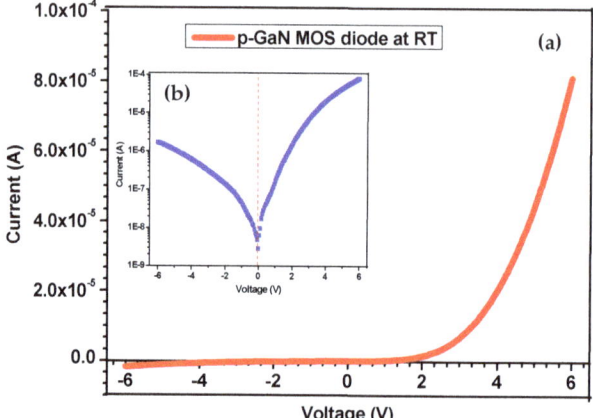

Figure 3. (a) I–V plot of the as-deposited Al/SiO$_2$/p-GaN MOS Schottky diode tested at RT, (b) the forward and reverse lnI–V characteristics of diodes.

According to the thermionic emission (TE) mode (for $qV > 3\,kT$), the electrical properties of the Schottky diode can be described as [6,11,22]:

$$I = I_0 \exp[q(V - IR_s)/nkT] \tag{1}$$

The SBH can be expressed by [1,5,21]:

$$\phi_B = \frac{kT}{q} \ln\left(\frac{AA^*T^2}{I_0}\right) \tag{2}$$

where I_0 is the saturation current, V is the applied voltage, R_s is the series resistance, n is the ideality factor, T is the measured temperature in Kelvin, q is the electronic charge, k is the Boltzmann constant, ϕ_B is the Schottky barrier height (SBH), A* is the Richardson constant, A is the contact area of the diode, and A* is the effective. The saturation current I_0 was defined by the intersection between the interpolated straight lines of the linear region and the current axis.

Using a stainless-steel mask with a square opening, the electrodes of our diode were measured at 1 mm^2. The A* value was 26.4 A·cm^{-2}·K^{-2} (based on effective mass $m^* = 0.22 \times m_e$ for GaN, m_e is electron mass) [4,5,13]. The ideality factor (n) from Equation (1) can be determined by [5,21,23]:

$$n = \frac{q}{kT}\left(\frac{dV}{d(\ln I)}\right) \tag{3}$$

Based upon Equations (1) and (2), the SBH of the diode was 0.67 V, while the ideality factor n, based on Equation (3), was 3.32. According to Cheung's method, the series resistance R_s and ideality factor can be found by the intersecting slope from the linear region of the dV/d(lnI) vs. the I plots [11,22,24–26]:

$$\frac{dV}{d(\ln I)} = \frac{nkT}{q} + IR_S \tag{4}$$

As shown in Figure 4, a calculation based on Equation (4) showed that the values of R_s and n were 5914 Ω and 3.51, respectively. Our MOS Schottky diode had high series resistance because there was an insulator SiO$_2$ layer of 70 nm between the metal and semiconductor.

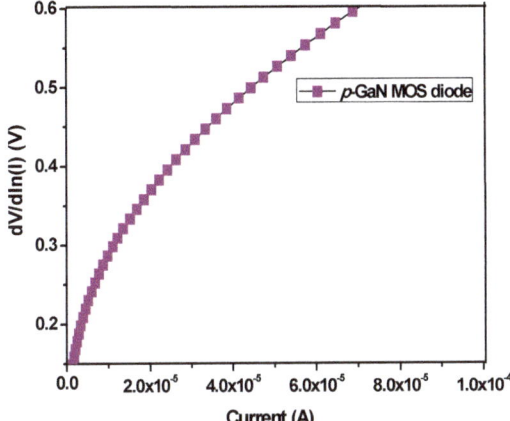

Figure 4. Plot of $dV/d\ln(I)$ versus I for the as-deposited Al/SiO$_2$/p-GaN MOS Schottky diode.

The Norde method was also used to calculate the effective SBH of the diodes. The Norde function is described as the $F(V, I)$ vs. the voltage V. It is given by [6,27]:

$$F(V,I) = \frac{V}{\gamma} - \frac{kT}{q}\ln\left(\frac{I}{AA^*T^2}\right) \quad (5)$$

The effective SBH Φ_B is obtained by:

$$\Phi_B = F(V_{min}) + \frac{V_{min}}{\gamma} - \frac{kT}{q} \quad (6)$$

where γ is the first integer (dimensionless) is higher than n, $F(V_{min})$ is the min value of $F(V)$, and V_{min} is the corresponding voltage [27,28].

Figure 5 displays the plot of $F(V)$ vs. the V of the Al/SiO$_2$/p-GaN MOS Schottky diode measured at RT. Based on Equations (5) and (6), the SBH value was 0.78 eV for the device. Table 1 lists all the parameters of the diode, calculated by I–V test, and Cheung's and Norde's methods.

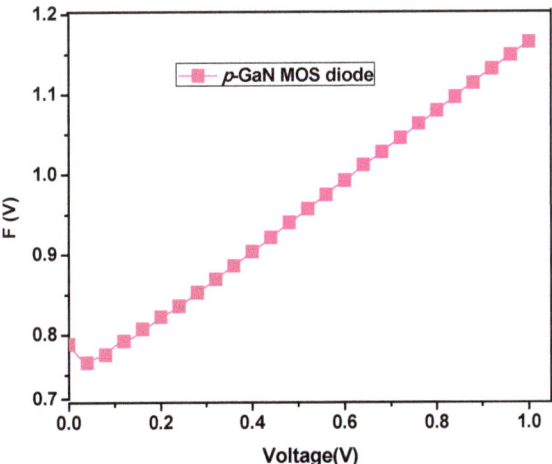

Figure 5. Characterization of the $F(V, I)$ vs. V for the as-deposited Al/SiO$_2$/p-GaN MOS Schottky diode.

Table 1. The parameters of the I–V characteristics of the Al/SiO₂/p-MOS Schottky diode at room temperature.

Sample	Leakage Current (A) at −1 (V)	Schottky Barrier Height (SBH) (eV)		From I–V	Cheungs' Function dV/dln(I)–I	
		I–V	Norde	n	R_s (Ω)	n
As-dep.	4.49×10^{-8}	0.67	0.78	3.32	5914	3.51

3.3. Capacitance–Voltage (C–V) Characteristics

The capacitance–voltage (C–V) measurement of our diode was expressed and tested at room temperature, with a frequency range of 100 kHz–1 MHz. The C–V relationship of diodes can be expressed by [4,9,13]:

$$\frac{1}{C^2} = \frac{2(V_{bi} - \frac{kT}{q} - V)}{q\varepsilon_s N_p A^2} \tag{7}$$

$$N_p = \frac{2}{q\varepsilon_s A^2}\left[-\frac{1}{d(1/C^{-2})/dV}\right] \tag{8}$$

where N_p is hole concentration, V is the flat band voltage, A is the area of the diode, and ε_s is the permittivity of the semiconductor ($\varepsilon_s = 9.5 \times \varepsilon_o$ for GaN, ε_o is electric constant) [4,13]. V_0 is determined by the plot of $1/C^2$ vs. V. The potential V_{bi} is calculated from V_0 by [4,5,11]:

$$V_{bi} = V_0 + \frac{kT}{q} \tag{9}$$

The SBH ϕ_{CV} from the C–V measurement is given by [13,14,18]:

$$\phi_{CV} = V_{bi} + E_F - \Delta\Phi_b \tag{10}$$

where E_F is the energy of Fermi level. This is given by [9,11,13]:

$$E_F = \frac{kT}{q}\ln(\frac{N_c}{N_p}) \tag{11}$$

Based on the $m^* = 0.22 \times m_e$ for GaN, N_c is the density of states in the conduction band edge. It is expressed by [2,5,7]:

$$N_c = 2\left(\frac{2\pi m^* kT}{h^2}\right)^{3/2} \tag{12}$$

where h is Plank constant. The $\Delta\Phi_b$ is the image force-induced barrier lowering. It is given by [7,9,13]:

$$\Delta\Phi_b = \left[\frac{qE_m}{4\pi\varepsilon_s\varepsilon_0}\right]^{1/2} \tag{13}$$

where E_m is the maximum electric field and given by [9,13]:

$$E_m = \left[\frac{2qN_p V_0}{\varepsilon_s\varepsilon_0}\right]^{1/2} \tag{14}$$

Figure 6a shows the plotted C–V measurement of the MOS Schottky diode tested at the frequency range 100 KHz–1 MHz. The Figure 6b is the electrical properties of the diode, which was measured at a frequency of 1 MHz with an alternating current (AC) modulation of 100 mV. Figure 7 shows the characterization of $1/C^2$ vs. V as a function of the p-MOS Schottky diode tested at different frequencies.

The *x*-intercept of the $1/C^2$ vs. V plot determined V_0 from the straight lines for the downward region at the reverse bias [4,5,7,11].

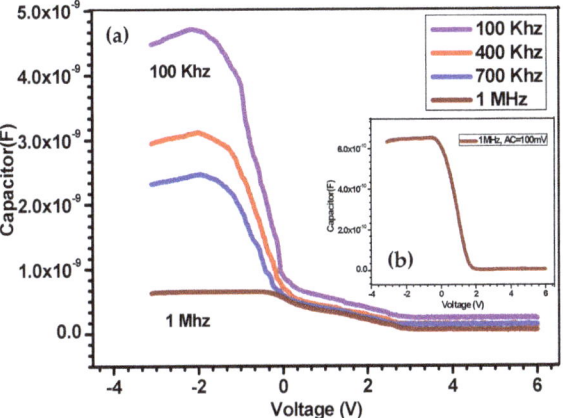

Figure 6. (**a**) Plot *C*–*V* measurement for the as-deposited Al/SiO$_2$/*p*-GaN MOS Schottky diode measurement at different frequencies between 100 kHz and 1 MHz, AC = 100 mV. (**b**) The electrical properties of the diode was measured at frequency of 1 MHz, AC = 100 mV.

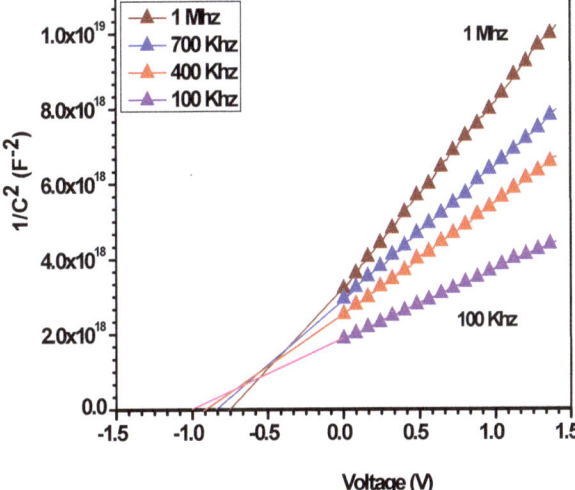

Figure 7. Plot of the $1/C^{-2}$–V for the as-deposited Al/SiO$_2$/*p*-GaN MOS Schottky diode measurement at different frequencies between 100 kHz and 1 MHz, AC = 100 mV.

Based on Equation (8), the hole concentration of the diode increased from 3.92×10^{16} cm^{-3} to 5.36×10^{16} cm^{-3} tested at 100 KHz to 1 MHz, while the Fermi level of the diode slightly decreased from 0.109 to 0.099 eV. After calculating the values of V_0, V_{bi}, E_F, E_m, and $\Delta\Phi_b$, based on Equations (9)–(14), the values of V_0 and the SBH values were reduced from 0.99 to 0.75 eV and 1.06 to 0.88 eV, respectively, when the testing frequencies were changed from 100 kHz to 1 MHz (Table 2).

Table 2. The parameters calculated from $1/C^2-V$ for characteristics of the Al/SiO$_2$/p-GaN MOS Schottky diode between 100 kHz and 1 MHz at the room temperature.

Frequency (KHz)	N_p (cm^{-3})	E_F (eV)	V_0 (eV)	$\Delta\Phi_b$ (eV)	Φ_{CV} (eV)
1000	3.92×10^{16}	0.109	0.75	0.039	0.88
700	4.20×10^{16}	0.106	0.82	0.042	0.94
400	4.89×10^{16}	0.103	0.90	0.044	1.02
100	5.36×10^{16}	0.099	0.99	0.046	1.06

4. Discussion

Tables 1 and 2 display all the parameters of the MOS Schottky diode measured by I–V and C–V measurements. This MOS Schottky diode showed an acceptable leakage current of ~4.49×10^{-8} A at -1 V, a breakdown voltage of -6 V, and a turn-on voltage of ~2.1 V. Calculating using Cheung's method, the MOS Schottky had R_s of 5918 Ω and n of 3.51. In addition, from the plot of the I–V curve, the ideality factor n was found to be 3.32. This indicated that a higher turn-on voltage leads to a higher ideality factor [4,9,13]. The growth of an insulator SiO$_2$ layer can effect to the accumulation layer during the forward bias, which affected to the high value of R_s in our Schotky diode. For similar results, by measurement at RT, the series resistance (R_s) and ideality factor (n) were calculated to be 84.4 kΩ and 2.96 for the Pt/SiO$_2$/n-GaN MOS Schottky diode [6], 230 Ω and 1.6 for the Au/SiO$_2$/n-GaN MOS diode [13], and 66 Ω and 2.48 for the Al/SnO$_2$/p-Si (111) MOS diode, respectively [18].

From Table 2, we showed that the C–V measurement data depended strongly on the tested frequency at the RT. At the high frequency (1 MHz), the interface state density could not identify the value of capacitance because it balanced with the semiconductor. At the low frequency, the interface state's density easily followed the AC signal. This created a signal and extra capacitance [11,13,18]. The N_p and SBH decreased from 5.36×10^{16} cm^3 to 3.92×10^{16} cm^3 and from 1.06 to 0.88 eV, respectively, with increased frequency, due to the existence of the interfacial SiO$_2$ layer in the depletion region. Similarly, the N_p of our Schottky diode was 5.36×10^{16} cm^{-3} when tested at a frequency of 1 MHz at RT [7]. The N_p and SBH of the Au/SiO$_2$/n-GaN MOS diode were 2.08×10^{17} cm^{-3} and 0.99 eV, also tested at a frequency of 1 MHz [9].

The hole's concentration (N_p), calculated from the $1/C^2$–V plots for the MOS Schottky diode, was lower than that calculated by the Hall measurement because this result was measured from the p-Mg-GaN film deposited on the SiO$_2$/Si substrate. In addition, with the fast growth rate in deposition, the interface between p-GaN and SiO$_2$ affected the polarized SiO$_2$ layer. The electrical field across the depletion region changed significantly near the p-GaN layer. It was affected by the strong variation in the hole's concentration, leakage current, and turn-on and breakdown voltages of the diode [11,15,16].

Therefore, the SBH of our MOS Schottky diode, in terms of C–V measurement, was higher than the I–V test because of the charges induced by dielectric layer. The SBH of our diode corresponded with some results (0.67–1.06 eV) of the GaN Schottky diodes made by metal organic chemical vapor deposition (MOCVD) and other approaches. Cheng et al. also reported the SBH of a Pt-oxide-Al$_{0.3}$Ga$_{0.7}$As MOS diode, which decreased from 1.03 to 0.86 eV after annealing in a hydrogen atmosphere [16]. Baris et al. reported all the parameters of Au/TiO/n-Si (100) MOS diodes; an ideality factor of 3.72 and an SBH of 0.62 eV were determined by testing the I–V measurement. Meanwhile, via calculation by C–V measurement, the SBH and bulk concentration were determined to be 0.99 eV and 9.82×10^{14} cm^{-3}, respectively [23].

5. Conclusions

The modeling and electrical properties of the Al/SiO$_2$/p-GaN MOS Schottky diode were successfully established by total RF sputtering. All the parameters were calculated based on I–V and C–V measurements. The SBHs of the MOS Schottky diodes were determined to be 0.67 (I–V), 0.78 eV (Norde), and 0.88 eV (C–V). The hole's concentration, tested by C–V measurement, decreased slightly compared with that determined by the Hall measurement. This was due to the existence of the SiO_2 layer in the p-GaN MOS diode. Our work using cost-effective RF sputtering to make the Al/SiO_2/p-GaN MOS Schottky diode can be applied to the development of electronic devices.

Author Contributions: Data curation, P.T.C. and T.A.T.T.; methodology, writing—original draft, investigation, P.T.C. and T.A.T.T., formal analysis, funding acquisition, writing—review and editing, P.T.C., T.A.T.T., P.Q.-P., V.K.N. and N.P.L.T.; supervision, D.-H.K.

Funding: This research was funded by the Ministry of Science and Technology of the Republic of China under grant number 107-2221-E-011-141-MY3.

Conflicts of Interest: The authors declare no conflict of interest.

References

1. Arslan, E.; Bütüna, S.; Safak, Y.; Özbaya, E.; Uslu, H. Electrical characterization of MS and MIS structures on AlGaN/AlN/GaN heterostructures. *Microelectron. Reliab.* **2011**, *51*, 370–375. [CrossRef]
2. Chang, P.C.; Yu, C.L.; Chiu, Y.J.; Liu, C.H.; Chang, S.J.; Su, Y.K.; Chuang, R.W. Ir/Pt Schottky contact oxidation for nitride-based Schottky barrier diodes. *Phys. Status Solidi C* **2007**, *5*, 1625–1628. [CrossRef]
3. Li, C.C.; Kuo, D.H. Effects of growth temperature on electrical and structural properties of sputtered GaN films with a cermet target. *J. Mater. Sci. Mater. Electron.* **2014**, *25*, 1404–1409. [CrossRef]
4. Reddy, M.S.; Reddy, V.R.; Choi, C.J. Electrical properties and interfacial reactions of rapidly annealed Ni/Ru Schottky rectifiers on n-type GaN. *J. Alloys. Comp.* **2010**, *503*, 186–191. [CrossRef]
5. Reddy, R.; Rao, P.K. ;Annealing temperature effect on electrical and structural properties of Cu/Au Schottky contacts to n-type GaN. *Microelectron. Eng.* **2008**, *85*, 470–476. [CrossRef]
6. Tuan, T.T.A.; Kuo, D.H.; Li, C.C.; Yen, W.C. Schottky barrier characteristics of Pt contacts to all sputtering-made n-type GaN and MOS diodes. *J. Mater. Sci. Mater. Electron.* **2014**, *25*, 3264–3270. [CrossRef]
7. Bengi, S.; Bülbül, M.M. Electrical and dielectric properties of Al/HfO_2/p-Si MOS device at high temperatures. *Curr. Appl. Phys.* **2013**, *25*, 1819–1825. [CrossRef]
8. Tsevas, S.; Vasilopoulou, M.; Kouvatsos, D.N.; Speliotis, T.; Niarchos, D. Characteristics of MOS diodes fabricated using sputter-deposited W or Cu/W film. *Microelectron. Eng.* **2006**, *83*, 1434–1437. [CrossRef]
9. Reddy, V.R.; Reddy, M.S.; Lakshmi, B.P.; Kumar, A. Electrical characterization of Au/n-GaN metal-semiconductor and Au/SiO_2/n-GaN metal-insulator-semiconductor structures. *J. Alloys. Comp.* **2011**, *509*, 8001–8007. [CrossRef]
10. Lin, K.W.; Chen, H.; Cheng, C.C.; Chuang, H.-M.; Lu, C.T.; Liu, W.C. Characteristics of a new Pt/oxide/$In_{0.49}Ga_{0.51}$P hydrogen-sensing Schottky diode. *Sens. Act. B* **2003**, *94*, 145–151. [CrossRef]
11. Tuan, T.T.A.; Kuo, D.H. Characteristics of RF reactive sputter-deposited Pt/SiO_2/n-InGaN MOS Schottky diodes. *Mater. Sci. Semicond. Process.* **2015**, *30*, 314–320. [CrossRef]
12. Liu, I.P.; Chang, C.H.; Huang, J.; Lin, K.W. Hydrogen sensing characteristics of a Pd/Nickel oxide/GaN-based Schottky diode. *Int. J. Hydrogen Energy* **2019**, *44*, 5748–5754. [CrossRef]
13. Lakshmi, B.P.; Reddy, M.S.; Kumar, A.A.; Reddy, V.R. Electrical transport properties of Au/SiO_2/n-GaN MIS structure in a wide temperature range. *Curr. Appl. Phys.* **2012**, *12*, 765–772. [CrossRef]
14. Juang, F.R.; Fang, Y.K.; Chiu, H.Y. Dependence of the Au/SnO_x/n-LTPS/glass thin film MOS Schottky diode CO gas sensing performances on operating temperature. *Microelectron. Reliab.* **2012**, *32*, 160–165. [CrossRef]
15. Tsai, T.H.; Huang, J.R.; Lin, K.W.; Hsu, W.C.; Chen, H.I.; Liu, W.C. Improved hydrogen sensing characteristics of a Pt/SiO_2/GaN Schottky diode. *Sens. Act. B* **2008**, *129*, 292–302. [CrossRef]
16. Cheng, C.C.; Tsai, Y.Y.; Lin, K.W.; Chen, H.I.; Lu, C.T.; Liu, W.C. Hydrogen sensing characteristics of a Pt-oxide-$Al_{0.3}Ga_{0.7}$As MOS Schottky diode. *Sens. Act. B* **2004**, *99*, 292–302. [CrossRef]
17. Chen, H.I.; Chang, C.C.; Lu, H.H.; Liu, I.-P.; Chen, W.C.; Ke, B.Y.; Liu, W.C. Hydrogen sensing performance of a Pd/HfO_2/GaN metal-oxide-semiconductor (MOS) Schottky diode. *Sens. Act B* **2018**, *262*, 852–859. [CrossRef]
18. Karadeniz, S.; Tuğluoğlu, N.; Serin, T. Substrate temperature dependence of series resistance in Al/SnO_2/p-Si (111) Schottky diodes prepared by spray deposition method. *Appl. Surf. Sci.* **2004**, *233*, 5–13. [CrossRef]
19. Chang, C.F.; STai, T.H.; Chen, H.I.; Lin, K.W.; Chen, T.P.; Chen, L.Y.; Liu, Y.C.; Liu, W.C. Hydrogen sensing properties of a Pd/SiO_2/AlGaN-based MOS diode. *Electrochem. Communi.* **2009**, *11*, 65–67. [CrossRef]

20. Li, C.C.; Kuo, D.H. Material and technology developments of the totally sputtering-made *p-n* GaN diodes for cost-effective power electronics. *J. Mater. Sci. Mater. Electron.* **2014**, *25*, 1942–1948. [CrossRef]
21. Tuan, T.T.A.; Kuo, D.H.; Li, C.C.; Li, G.Z. Effect of temperature dependence on electrical characterization of *p-n* GaN diode fabricated by RF magnetron sputtering. *Mater. Sci. Appl.* **2015**, *6*, 809–817.
22. Ramesh, C.K.; Reddy, V.R.; Choi, C.J. Electrical characteristics of molybdenum Schottky contacts on *n*-type GaN. *Mater. Sci. Eng. B* **2004**, *112*, 30–33. [CrossRef]
23. Barış, B. Analysis of device parameters for Au/TiO$_2$/*n*-Si (100) metal-oxide semiconductor (MOS) diodes. *Phys. B* **2014**, *438*, 65–69. [CrossRef]
24. Cheung, S.K.; Cheung, N.W. Extraction of Schottky diode parameters from forward current-voltage characteristics. *Appl. Phys. Lett.* **1986**, *49*, 85–87. [CrossRef]
25. Tuan, T.T.A.; Kuo, D.H.; Saragih, A.D.; Li, G.Z. Electrical properties of RF-sputtered Zn-doped GaN films and *p*-Zn-GaN/*n*-Si hetero junction diode with low leakage current of 10^9 A and a high rectification ratio above 10^5. *Mater. Sci. Eng. B* **2017**, *222*, 18–25. [CrossRef]
26. Ravinandan, M.; Rao, P.K.; Reddy, V.R. Temperature dependence of current-voltage *I–V* characteristics of Pt/Au Schottky contacts on n-type GaN. *J. Optoelectron. Adv. Mater.* **2008**, *10*, 2787–2792.
27. Norde, H. A modified forward IV plot for Schottky diodes with high series resistance. *J. Appl. Phys.* **1979**, *50*, 5052–5053. [CrossRef]
28. Tuan, T.T.A.; Kuo, D.H. Temperature-dependent electrical propertiesof the sputtering-made *n*-InGaN/*p*-GaN junction diode with a breakdown voltage above 20 V. *Mater. Sci. Semicond. Process.* **2015**, *32*, 160–165. [CrossRef]

© 2019 by the authors. Licensee MDPI, Basel, Switzerland. This article is an open access article distributed under the terms and conditions of the Creative Commons Attribution (CC BY) license (http://creativecommons.org/licenses/by/4.0/).

Article

Electrical Characterization of RF Reactive Sputtered p–Mg-In$_x$Ga$_{1-x}$N/n–Si Hetero-Junction Diodes without Using Buffer Layer

Thi Tran Anh Tuan [1], Dong-Hau Kuo [2,*], Phuong Thao Cao [3,*], Van Sau Nguyen [1], Quoc-Phong Pham [3], Vinh Khanh Nghi [3] and Nguyen Phuong Lan Tran [4]

1. School of Basic Sciences, Tra Vinh University, Tra Vinh 87000, Vietnam; thitrananhtuan@tvu.edu.vn (T.T.A.T.); nvsau@tvu.edu.vn (V.S.N.)
2. Department of Materials Science and Engineering, National Taiwan University of Science and Technology, Taipei 10607, Taiwan
3. School of Engineering and Technology, Tra Vinh University, Tra Vinh 87000, Vietnam; phongpham@tvu.edu.vn (Q.-P.P.); nghivinhkhanh@tvu.edu.vn (V.K.N.)
4. College of Engineering and Technology, Can Tho University, Can Tho 94000, Vietnam; tnplan@ctu.edu.vn
* Correspondence: dhkuo@mail.ntust.edu.tw (D.-H.K.); cpthao@tvu.edu.vn (P.T.C.); Tel.: +886-2-27303291 (D.-H.K.)

Received: 30 August 2019; Accepted: 21 October 2019; Published: 25 October 2019

Abstract: The modeling of p–In$_x$Ga$_{1-x}$N/n–Si hetero junction diodes without using the buffer layer were investigated with the "top-top" electrode. The p–Mg-GaN and p–Mg-In$_{0.05}$Ga$_{0.95}$N were deposited directly on the n–Si (100) wafer by the RF reactive sputtering at 400 °C with single cermet targets. Al and Pt with the square size of 1 mm^2 were used for electrodes of p–In$_x$Ga$_{1-x}$N/n–Si diodes. Both devices had been designed to prove the p-type performance of 10% Mg-doped in GaN and InGaN films. By Hall measurement at the room temperature (RT), the holes concentration and mobility were determined to be $N_p = 3.45 \times 10^{16}$ cm^{-3} and $\mu = 145$ cm^2/V·s for p–GaN film, $N_p = 2.53 \times 10^{17}$ cm^{-3}, and $\mu = 45$ cm^2/V·s for p–InGaN film. By the I–V measurement at RT, the leakage currents at −5 V and turn-on voltages were found to be 9.31×10^{-7} A and 2.4 V for p–GaN/n–Si and 3.38×10^{-6} A and 1.5 V for p–InGaN/n–Si diode. The current densities at the forward bias of 20 V were 0.421 and 0.814 A·cm^{-2} for p–GaN/n–Si and p–InGaN/n–Si devices. The electrical properties were measured at the temperature range of 25 to 150 °C. By calculating based on the TE mode, Cheungs' and Norde methods, and other parameters of diodes were also determined and compared.

Keywords: p–Mg-InGaN films; RF sputtering; I–V measurement; Cheung's method; Norde's method; TE mode

1. Introduction

GaN and InGaN have excellent characteristics such as high conductivity and high mobility. The development and creation of p–layer GaN and InGaN materials involve one of the important technologies in designing electronic devices [1–5]. The investigation of high-quality doping in GaN and InGaN semiconductors by incorporating elements such as Zn, and Cu, Mg for p–GaN behavior, and its alloys had reported [4–8]. The success of Mg doping in forming p–In$_x$Ga$_{1-x}$N films is an important factor for developing electric devices, a photo detector, and solar cell devices [7–10]. Si wafer has often been used for the growth of GaN, InGaN, and their alloys for applications in photo-detector, solar cells, and electronic devices. The combination between n–InGaN layers and n–Si wafers were studied to improve the interface layers by using an assortment of approaches [11–14]. For fabricated electronic devices, the growth of the InGaN layer on n–Si wafers were studied to improve the interface layers by using a variety of approaches [15–20]. Lee at al. studied electrical properties of a nanowire n–GaN/p–Si device by forming dielectrophoretic alignment. At the current density of 10–60 A/cm^2, the diode was well-defined with a forward voltage drop of 1.2–2.0 V and high resistance in the range of 447 KΩ [17]. Li et al. used the RF sputtering method to deposit the Mg-doped GaN films on the Si substrate and design the homo junction GaN diodes. By testing the I–V measurement, the turn-on voltages of diodes were 2.3 and 2.1 V for as-deposited and 500 °C-annealed sample, respectively [21]. Vinay Kabra et al. investigated the p–ZnO/n–Si hetero junction diode by using a dip coating technique. Their electrical properties showed the highly rectifying, with a rectification ratio of 101 at 3 V [22]. Mohd Yusoff et al. reported the p–n junction diode based on GaN grown on the AlN/Si (111) substrate and annealed samples at 700 °C. The ideality factors of their diodes decreased from 19.68 to 15.14, with the testing temperature increasing from 30 to 104 °C [23]. In previous work, the Mg–doped $In_xGa_{1-x}N$ films had been deposited on Si (100) substrates by RF reactive sputtering. The Mg–$In_xGa_{1-x}N$ films had the p–type conduction at $x \leq 0.075$. The p–Mg–$In_{0.05}Ga_{0.95}N$/n–GaN diode was shown the leakage current of 2.7×10^{-6} A, turn-on voltage of 1.8 V, and breakdown voltage of 6.8 V at the RT [12].

All previous groups often investigated the p–GaN films and their alloys by using MOCVD above 800 °C and other methods. The sputtered technique with the low temperature at and below 400 °C has been hardly declared. Furthermore, there is no report on the electrical properties of p–Mg-$In_xGa_{1-x}N$/n–Si diodes without using buffer layers. In this study, to prove the success of Mg doping in the p–GaN and the p–InGaN films, the modeling of p–Mg-$In_xGa_{1-x}N$/n–Si hetero junction diodes was designed by using the RF reactive sputtering. This method was chosen to design diodes due to the benefits of low sputtered-temperature, low cost, and safe working atmosphere [7,13,15]. The n–Si (100) wafer was also used for its low cost, large wafer size, and easy availability [18,22,24]. The electrical characteristics of devices were calculated by the thermionic emission (TE) mode at different testing temperatures [3,12,13].

2. Materials and Methods

Figure 1 shows the structural modeling of a p–$In_xGa_{1-x}N$/n–Si hetero junction diode. This device was designed on the n–Si (100) wafer by modeling with the "top-top" electrode. The p–Mg-$In_{0.05}Ga_{0.95}N$ and p–Mg-GaN films were deposited together on the n–Si (100) and SiO_2/Si (100) substrate. The n–Si (100) wafer had sheet resistance of ~1–10 Ω·cm, diameter of 2 inches, thickness of ~550 μm, and the polished surface. Sputtered-$In_xGa_{1-x}N$ films on SiO_2/Si substrate were used for testing Hall measurement and SEM analysis.

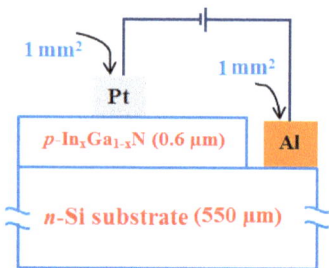

Figure 1. Structural modeling of p–$In_xGa_{1-x}N$/n–Si hetero junction diode.

For the p–GaN film, the sputtering cermet target was made by hot pressing with the mixture of metallic Mg, Ga powders, and GaN powder. The [Mg]/([Ga]+[Mg]) molar ratio in each cermet target was reserved at 10%. Similar for p–InGaN film, the [Mg]/([In]+[Ga]+[Mg]) molar ratio in each cermet target was also kept at 10%. The [In]/([In]+[Ga]+[Mg]) molar ratios was 5%. The p–Mg-$In_xGa_{1-x}N$ films (with x~0 and 0.05) were deposited on n–Si (100) substrates at 400 °C for 25 min by RF reactive sputtering. Both Mg–$In_xGa_{1-x}N$ targets were sputtered with RF power of 150W under the gas mixture of Ar and N_2, which remained at 5 sccm for each.

The pure Aluminum (99.99%) and Platinum (99.99%) targets were used for making the electrodes for p–In$_x$Ga$_{1-x}$N/n–Si hetero junction diodes. By using steel masks with the square size of 1 mm^2, these electrodes were deposited at 200 °C, with RF power of 80W for 30 min on top of p–In$_x$Ga$_{1-x}$N films and n–Si substrate. The detail procedure for creating the p–GaN and p–InGaN films, made by RF reactive sputtering, can be referred to our previous works [10–12,21].

In this work, our diodes were designed at the low sputtered-temperature and high pressure. The holes' concentration, electrical conductivities, and mobilities of p–Mg-In$_x$Ga$_{1-x}$N films and n–Si wafer were measured by a Hall measurement (HMS–2000, Ecopia, Tokyo, Japan). Scanning electron microscopy (SEM, JSM–6500F, JEOL, Tokyo, Japan) was used to observe the surface morphology of p–In$_x$Ga$_{1-x}$N films. Energy dispersive spectrometer (EDS, JSM–6500F) equipped on SEM was used for composition analysis of films. The I–V characteristics of p–In$_x$Ga$_{1-x}$N/n–Si diodes were measured by using Semiconductor Device Analyzer (Agilent, B1500A, Tokyo, Japan) with the different temperature. All parameters of diodes were calculated by following the equations of the TE mode as well as the Cheungs' and Norde method.

3. Results and Discussion

3.1. Structural and Electrical Characteristics

The carrier concentration (N_p) and carrier mobility (μ) of n–Si wafer were found to be $N_p = 4.7 \times 10^{15}$ cm^{-3} and = 196 cm^2/V.s when measuring the Hall effect at room temperature (RT). The holes concentration (N_p) and mobility (μ) were also determined to be $N_p = 3.45 \times 10^{16}$ cm^{-3}, $\mu = 145$ cm^2/V·s for p–GaN film, and $N_p = 2.53 \times 10^{17}$ cm^{-3}, $\mu = 45$ cm^2/V·s for p–InGaN film, respectively. By SEM analysis, the thicknesses of Pt and Al layers were ~250 nm and the thickness of both p–In$_x$Ga$_{1-x}$N films was ~0.6 µm.

3.2. The Energy Band Diagram

Figure 2 displays the energy band diagram of a p–In$_{0.5}$Ga$_{0.95}$N/n–Si diode as an example. Our band gap (E_g) of Mg-doped In$_{0.5}$Ga$_{0.95}$N (Mg of 10%) films at 400 °C-deposition were 2.92 eV. The E_g was reduced by increasing In content in In$_x$Ga$_{1-x}$N [10,21]. In a similar result, Wagner et al. reported the composition dependence of the band gap energy of the strained In$_x$Ga$_{1-x}$N layers grown by MOCVD. The energy E_g of In$_x$Ga$_{1-x}$N decreased from 3.43 to 3.28 eV with the increasing in ratio from 0.02 to 0.15 [25]. The band gap values of n–Si was also found to be 1.12 eV. The electron affinities χ of p–InGaN and n–Si were 4.20 eV and 4.05 eV, respectively [6,24]. According to the band gap values of p–In$_{0.5}$Ga$_{0.95}$N and n–Si (2.92 and 1.12 eV), the barrier ΔE_c for electrons was followed by $\Delta E_c = \chi_{p-InGaN} - \chi_{Si}$ and the barrier ΔE_v for holes was calculated by $\Delta E_v = E_{g(InGaN)} - E_{g(Si)} + \Delta E_c$ (Figure 2). As a result, the energy band diagram showed a small conduction band offset of 0.15 eV and a large valance band offset of 1.95 eV [22,24–26]. Therefore, the electrons' injection from n–Si into p–In$_x$Ga$_{1-x}$N will be easier than the holes' injection from p–In$_x$Ga$_{1-x}$N into n–Si because the energetic barrier ΔE_v of holes is many times higher than the barrier ΔE_c of electrons. With the existing space charge layer at the forward bias, the current is limited. However, the depletion width will decrease because of the low potential barrier, and the current will increase easily by increasing the voltage [6,27,28].

3.3. I–V Measurements

Figure 3 shows the current density–voltage characteristics of p–GaN/n-Si (device-A) and p–InGaN/n-Si (device-B) tested at RT based on a 1-mm^2 contact. The applied voltage of measurement was extended from −20 V to +20 V for both devices. The current densities and the leakage current densities are found in the area of 1 cm^2.

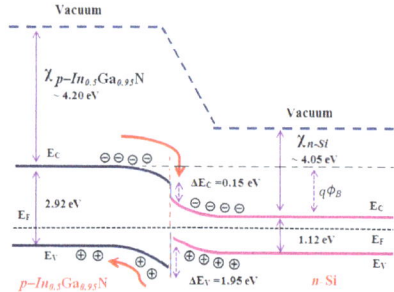

Figure 2. Energy band diagram of the p–In$_{0.5}$Ga$_{0.95}$N/n–Si hetero junction diode.

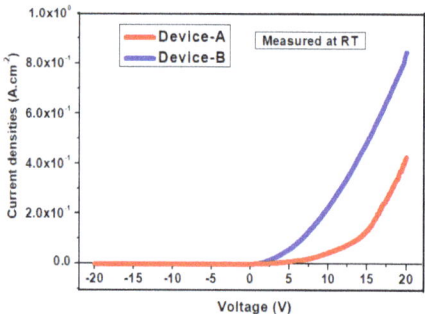

Figure 3. Current–voltage characteristics of p–GaN/n-Si and p–InGaN/n-Si diode, measured at room temperature.

At the bias range of (−5 V, +5 V), the leakage current of diodes at −1 V and the turn-on voltages at RT were 1.43 × 10^{-7} A, 1.5 V (device-A) and 2.58 × 10^{-7} A, 2.4 V (device-B). Our devices had an improvement in the turn-on voltage with the increase in content in the p–InGaN layer.

With the wide voltage of (−20 V, +20 V), Figure 3 also shows the breakdown voltage was beyond the largest instrument capacity of ~20 V for both diodes. The leakage currents also were found to be 9.31 × 10^{-7} and 3.38 × 10^{-6} A at −5 V for (device-A) and (device-B), respectively. At the forward bias of 20 V, the current densities of diodes were 0.421 and 0.814 A·cm^{-2} for device-A and device-B, respectively. Higher holes concentration in the p–InGaN layer with the performance of the in ratio in p–InGaN leads to the higher current density of device-B. In addition, with the high conduction of electrodes, the forward current of diodes increased rapidly at the bias of (−20 V, +20 V). The rectification ratio (on/off) of diodes were also found to be 3.22 × 10^4 (device-A) and 6.61 × 10^4 (device-B), respectively.

Figure 4 shows the semilogarithmic I–V characteristics of diodes under the forward bias, which were tested at the RT temperature. All parameters of diodes can be calculated by the thermionic-emission (TE) mode (for $qV > 3\,kT$). They are given by the equations below [3,11,12].

$$I = I_0 \left[\exp \frac{q}{nkT}(V - IR_s) \right] \tag{1}$$

$$\Phi_b = \frac{kT}{q} \ln \left(\frac{AA^*T^2}{I_0} \right) \tag{2}$$

$$n = \frac{q}{kT} \left(\frac{dV}{d\ln I} \right) \tag{3}$$

$$A^* = \frac{4\pi q k^2 m^*}{h^3} \tag{4}$$

where A^* is the effective Richardson constant, R_s is the series resistance, n is the ideality factor, I_0 is the saturation current, q is the electronic charge, A is the area of diode, ϕ_b is the barrier height, m_e is the free electron mass, m^* is the effective electron mass, and h is the Plank constant [29–31]. In base Equation (4), the theoretical value of A^* was 26.4 A·cm^{-2}·K^{-2} for p–GaN ($m^* = 0.22m_e$), and 23 A·cm^{-2}·K^{-2} for p–InGaN ($m^* = 0.19m_e$) [6,12,13].

As shown in Figure 4, the barrier height of diodes can be calculated from I_0 as the saturation current. Based on Equation (2), and from the plot of ln$I$ versus V, the saturation current I_0 can be determined by intersecting the interpolated straight lines from the linear region with the current axis [12,20].

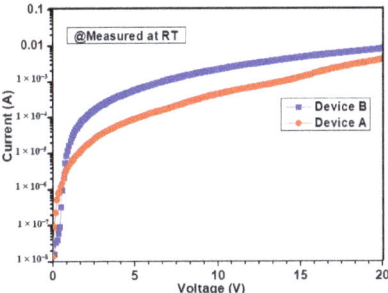

Figure 4. The I–V characteristics of device-A and device-A under the forward bias tested at room temperature.

Figure 5a,b show the semi logarithmic I–V characteristics of diodes were tested at the temperature range from 25 to 150 °C. From Figure 5, the leakage current of diode at −5 V increased from 9.39 × 10^{-7} A (device-A) and 3.38 × 10^{-6} (device-B) at 25 °C to 3.25 × 10^{-5} A (device-A) and 2.64 × 10^{-4} A (device-B) at 150 °C. Based upon Equations (1)–(3), the barrier height values were found to increase from 0.54 to 0.69 eV (device-A), and 0.50 to 0.62 eV (device-B), whereas the ideality factor n decreased from 5.9 to 4.8 (device-A), and 5.1 to 3.6 (devices-B) with testing temperatures ranging from 25 to 150 °C. These achieved parameters of diodes made by RF sputtering are similar to some previous works with the GaN-diodes made by MOCVD and techniques [1,3,18,28].

Similar to the analysis of the parameters for Schottky diodes, the effect of series resistance R_s will influence and change the ideality factor n. They can be determined by using Cheung's method. It is given by the equation below [13,32,33].

$$\frac{dV}{d(\ln I)} = \frac{nkT}{q} + IR_S \quad (5)$$

Based on Equation (5) and the plot of the $dV/d(\ln I)$ versus I from Figure 6, the series resistance R_s and ideality n can be calculated the from the the the slope and the intercept [11,33,34]. Table 1 shows all detailed parameters of the I–V measurement for both devices tested at different temperatures. With the high Indium content, the R_s and n values of device-B were smaller than device-A. They showed 412 Ω and 5.3 at 25 °C and decreased them to 133 Ω and 3.8 at 150 °C.

By comparison, with Cheung's method, the modified Norde function was also used to define the effective barrier height of diodes. It is followed by the equation below [12,13,34–36].

$$F(V, I) = \frac{V}{\gamma} - \frac{kT}{q} \ln(\frac{I}{AA^*T^2}) \quad (6)$$

The effective barrier height ϕ_B is followed by the equation below.

$$\phi_B = F(V_{min}) + \frac{V_{min}}{\gamma} - \frac{kT}{q} \quad (7)$$

where the F(V_{min}) is the minimum value of F(V,I), γ is the first integer (dimensionless) greater than n, and V_{min} is the corresponding voltage at F(V_{min}).

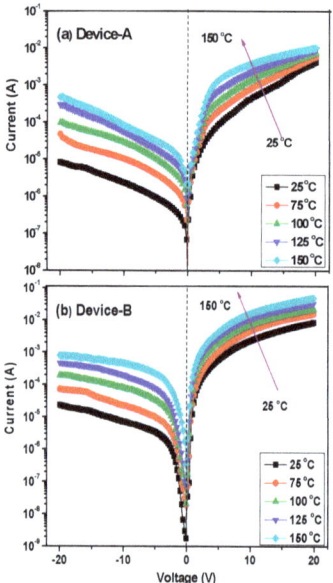

Figure 5. Forward and reverse (*I–V*) characteristics of (**a**) device-A and (**b**) device-B tested in the temperature range of 25–150 °C.

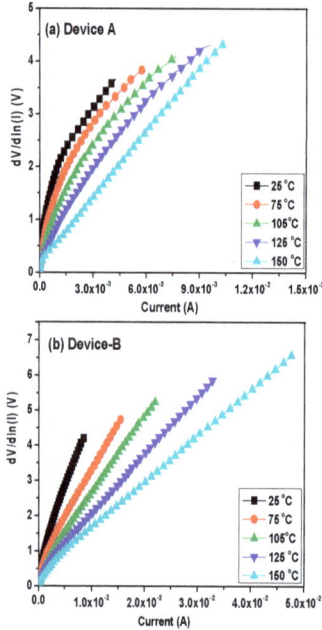

Figure 6. Plots of d*V*/dln(*I*) versus *I* for (**a**) device-A and (**b**) device-B tested in the temperature range of 25 to 150 °C.

Figure 7a,b shows the plot of F(V,I) versus V of hetero junction diodes as a function of the different temperatures tested. By using Equations (6) and (7), the barrier height values were determined to be 0.56 eV (device-A) and 0.53 eV (device-B) at 25 °C, and reached 0.71 eV (device A) and 0.64 eV (device B) at 150 °C (Table 1).

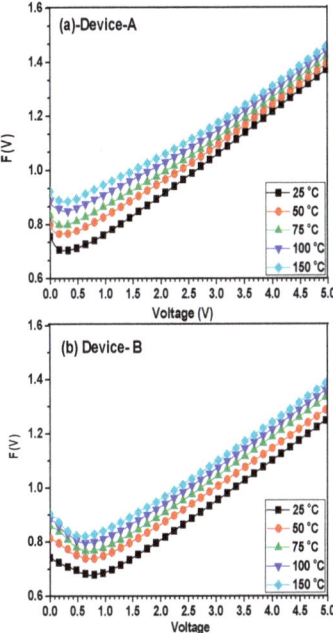

Figure 7. The $F(V)$ versus V plot for (**a**) device-A and (**b**) device-B tested in the temperature range of 25 to 150 °C.

4. Discussion

Table 1 showed the detailed comparisons of the I–V measurements for both hetero junction diodes. The smallest leakage current of 9.31×10^{-7} A at −5 V was found in device-A tested at RT, while the smallest turn-on voltage was ~1.5 V. The highest current density of 0.814 A·cm^{-2} were found in device-B at the forward bias of 20 V. At the reverse bias, the breakdown voltage of both devices was strongly blocked with the value at −20 V. The forward current increased quickly with the occurrence of the in content in the p–InGaN layer. With a higher holes carrier concentration of 2.53×10^{17} cm^{-3} and a smaller band gap of 2.92 eV, device-B showed the small turn-on voltage of 1.5 V, barrier heights of 0.5 eV (Cheung's method), and 0.53 eV (Norde method), compared with all parameters of device-A [20].

The ideality factor n of diodes decreased quickly, from 5.9 at 25 °C for device-A to 3.6 at 150 °C for device-B. For the Schottky diode, the turn-on voltage is often small, and the ideality factor n is <2 [2,3,13]. In our diodes, the ideality factors had high values because of the present high series' resistances, which were calculated from the slope of the linear region in Cheung's method. In addition, from the band diagram in Figure 2, the barrier ΔE_v for holes was larger than that of ΔE_c. Therefore, a higher valance band offset will have a high potential barrier, large turn-on voltage, high ideality factor, and high series resistance Rs. The high ideality factor of our diodes at RT also has been attributed to the high structural defect density, which serves as the trap-assisted generation-recombination centers. This procedure also influences the current transportation of hetero junction diodes [13,17,20]. The R_s values also reduced from 876 Ω at 25 °C (device-A) to 133 Ω at 150 °C (device-B). The generations of series resistance, interface states, and the voltage drop across the interfacial layer caused the slight difference between the ideality factor n calculated from the ln$I$ vs V and dV/d(ln$I$) vs. I plots (Table 1).

Both hetero junction diodes had remarkable improvements as compared with some other homo and hetero junction diodes made by different techniques. Their deposition had epitaxial growth and was conducted at high temperatures above 800 °C by MOVCD [18,19,27,37–39]. Our devices had displayed the special I–V characteristics, which can be attributed to the low-temperature deposited at −400 °C without using a buffer layer. Both our diodes showed the breakdown voltage above 20 V for a 1 mm^2-sized contact.

Table 1. The parameters calculated from electrical characteristics of hetero junction diodes as a function of testing temperatures.

Temp (°C)	Leakage Current (A) at −5 V	Barrier Height ϕ_B (eV) I–V	Norde	From I–V n	Cheung's Function dV/dln(I) versus I R_s (Ω)	n
Device-A						
25	9.31 × 10^{-7}	0.54	0.56	5.9	876	6.4
75	4.02 × 10^{-6}	0.57	0.58	5.6	683	5.8
100	1.03 × 10^{-5}	0.63	0.64	5.4	549	5.5
125	1.91 × 10^{-5}	0.66	0.68	5.0	467	5.3
150	3.26 × 10^{-5}	0.69	0.71	4.8	413	5.1
Device-B						
25	3.38 × 10^{-6}	0.50	0.53	5.1	486	5.3
75	9.92 × 10^{-5}	0.53	0.55	4.8	292	4.9
100	3.43 × 10^{-5}	0.57	0.60	4.6	233	4.7
125	7.81 × 10^{-5}	0.60	0.62	4.1	173	4.2
150	2.64 × 10^{-4}	0.62	0.64	3.6	133	3.8

K.E.F. Keskenler et al. made the n–ZnO/p–Si hetero junction diode by using a sol-gel spin technique [38]. The leakage current, barrier height, ideality factor, and series resistance of their diode were determined to be 4 × 10^{-6} A at −1 V, 0.71 eV, 2.03, and 42.1 Ω at RT. Chirakkara et al. investigated the n–ZnO/p–Si (100) diode by pulsed laser deposition without using a buffer layer [39]. By I–V testing at the temperature range of 300–390 K, the barrier height increased from 0.6 (300 K) to 0.76 eV (390 K). Seng et al. reported the n–ZnO/p–GaN diode formed by the RF technique [40]. Their diode had showed the turn-on voltage of 2 V, the leakage current of 1.6 × 10^{-5} A, and the series resistances of 102 Ω. Hsueh et al. investigated the n–Mg$_x$Zn$_{1-x}$O/p–GaN hetero-junction diode. Their ideality factors decreased from 3.86 to 7.00, with testing temperature increasing from 25 to 125 °C [41]. Compared to our previous works, for the sputtering-made n–In$_x$Ga$_{1-x}$N/p–Si devices with the similar area of 1-mm^2, the leakage current densities at −5 V were 5.96 × 10^{-5} A·cm^{-2} for n–GaN/p–Si and 2.81 × 10^{-4} A·cm^{-2} for n–In$_{0.4}$Ga$_{0.6}$N/p–Si, respectively [20].

At this time, there are rarely a report about the Mg-doped–In$_x$Ga$_{1-x}$N film made by RF sputtering below 400 °C deposition and designed diode on the n–Si wafer. With the improvement of doping Mg in In$_x$Ga$_{1-x}$N film, our devices had the stable I–V characteristics up to the testing temperature of 150 °C. The p–In$_x$Ga$_{1-x}$N/n–Si diodes with the strong breakdown voltage of 20 V and leakage current of ~10^{-7} A will create the opportunity to develop the power devices.

5. Conclusions

p–In$_x$Ga$_{1-x}$N/n–Si diodes were successfully investigated by directly depositing the p–GaN and p–Mg-doped-In$_{0.5}$Ga$_{0.95}$N films on n–Si(100) wafers without using a buffer layer. The highest current density of 0.814 A·cm^{-2} at 20 V and the smallest turn-on voltage of 1.5 V were found for p–InGaN/n–Si devices. Pt and Al for Omhic contacts contributed to the high current density at the forward bias of hetero junction diodes. Both diodes also displayed the good I–V characteristics up to the testing temperature of 150 °C and the breakdown voltages of 20 V. By calculating the equations based on the TE mode as well as Cheungs' and Norde method, the obtained electrical parameters were also compared. They can be referred to develop cost-effective solutions in electronic devices.

Author Contributions: Data Curation, P.T.C. and T.T.A.T.; Methodology, Writing—Original Draft, Investigation, P.T.C. and T.T.A.T.; Formal Analysis, Funding Acquisition, Writing—Review and Editing, P.T.C., T.T.A.T., V.S.N., Q.-P.P., V.K.N. and N.P.L.T.; Supervision, D.-H.K.

Funding: This research was funded by the Ministry of Science and Technology of the Republic of China under grant number 107-2221-E-011-141-MY3.

Conflicts of Interest: The authors declare no conflict of interest.

References

1. Baik, K.H.; Irokawa, Y.; Ren, F.; Pearton, S.J.; Park, S.S.; Park, S.J. Temperature dependence of forward current characteristics of GaN junction and Schottky rectifiers. *Solid State Electron.* **2003**, *47*, 1533–1538. [CrossRef]
2. Jang, J.S.; Kim, D.; Seong, T.Y. Schottky barrier characteristics of Pt contacts to n-type InGaN. *J. Appl. Phys.* **2006**, *99*, 073704. [CrossRef]
3. Reddy, V.R.; Prasanna, B.P.; Padma, R. Electrical Properties of Rapidly Annealed Ir and Ir/Au Schottky Contacts on n-Type InGaN. *J. Metall.* **2012**, *1*, 1–9. [CrossRef]
4. Cao, X.A.; Lachode, J.R.; Ren, F. Implanted p–n junction in GaN. *Solid State Electron.* **1999**, *43*, 1235–1238. [CrossRef]
5. Hickman, R.; Vanhove, J.M.; Chow, P.P.; Klaassen, J.J. GaN PN junction issues and developments. *Solid State Electron.* **2000**, *44*, 377–381. [CrossRef]
6. Tuan, T.T.A.; Kuo, D.H.; Albert, D.S.; Li, G.Z. Electrical properties of RF-sputtered Zn-doped GaN films and p-Zn-GaN/n -Si hetero junction diode with low leakage current of 10^{-9} A and a high rectification ratio above 10^5. *Mater. Sci. Eng. B* **2017**, *222*, 18–25. [CrossRef]
7. Islam, M.R.; Sugita, K.; Horie, M.; Islam, A.Y. Mg doping behavior of MOVPE $In_xGa_{1-x}N$ (x~0.4). *J.Cryst. Growth* **2009**, *311*, 2817–2820. [CrossRef]
8. Yohannes, K.; Kuo, D.H. Growth of p-type Cu-doped GaN films with magnetron sputtering at and below 400 °C. *Mater. Sci. Semicond. Process.* **2015**, *29*, 288–293.
9. Ager, J.W.; Miller, N.; Jones, R.E. Mg-doped InN and InGaN Photoluminescence, Capacitance–Voltage and thermo-power measurements. *Phys. Status Solidi* **2008**, *245*, 873–877. [CrossRef]
10. Kuo, D.H.; Li, C.C.; Tuan, T.T.A.; Yen, C.L. Effects of Mg Doping on the Performance of InGaN Films Made by Reactive Sputtering. *J. Electron. Mater.* **2014**, *44*, 210–216. [CrossRef]
11. Tuan, T.T.A.; Kuo, D.H.; Li, C.C.; Li, G.Z. Effect of temperature dependence on electrical characterization of p-n GaN diode fabricated by RF magnetron sputtering. *Mater. Sci. Appl.* **2015**, *6*, 809–817.
12. Tuan, T.T.A.; Kuo, D.H.; Li, C.C.; Yen, W.L. Electrical and structural properties of Mg-doped $In_xGa_{1-x}N$ ($x \leq 0.1$) and p-InGaN/n-GaN junction diode made all by RF reactive sputtering. *Mater. Sci. Eng. B* **2015**, *193*, 13–19.
13. Tuan, T.T.A.; Kuo, D.H.; Li, C.C.; Yen, W.L. Schottky barrier characteristics of Pt contacts to all sputtering-made n-type GaN and MOS diodes. *Mater. Sci. Mater. Electron.* **2014**, *25*, 3264–3270. [CrossRef]
14. Oh, M.; Lee, J.J.; Lee, L.K.; Oh, H.K. Electrical characteristics of Mg-doped p-GaN treated with the electrochemical potentiostatic activation method. *J. Alloy Compd.* **2014**, *585*, 414–417. [CrossRef]
15. Umeno, M.; Egawa, T.; Ishikawa., H. GaN-based optoelectronic devices on sapphire and Si substrates. *Mater. Sci. Semicond. Process.* **2001**, *4*, 459–466. [CrossRef]
16. Liou, B.W. Design and fabrication of $In_xGa_{1-x}N$/GaN solar cells with a multiple-quantum-well structure on SiCN/Si(111) substrates. *Thin Solid Films* **2011**, *520*, 1084–1090. [CrossRef]
17. Lee, S.Y.; Kim, T.H.; Suh, D.I.; Park, J.E. An electrical characterization of a hetero-junction nanowire (NW) PN diode (n-GaN NW/pSi) formed by dielectrophoresis alignment. *Phys. E* **2007**, *36*, 194–198. [CrossRef]
18. Zebbar, N.; Kheireddine, Y.; Mokeddema, K.; Hafdallah, A.; Kechouanea, M.; Aida, M.S. Structural optical and electrical properties of n-ZnO/p-Si heterojunction prepared by ultrasonic spray. *Mater. Sci. Semicond. Process.* **2011**, *14*, 229–234. [CrossRef]
19. Gad, A.E.; Hoffmann, M.W.G.; Hernandez, R.F. Coaxial p-Si/n-ZnO nanowire heterostructures for energy and sensing applications. *Mater. Chem. Phys.* **2012**, *135*, 618–622. [CrossRef]
20. Tuan, T.T.A.; Kuo, D.H.; Lin, K.; Li, G.Z. Temperature dependence of electrical characteristics of n-$In_xGa_{1-x}N$/ p-Si hetero-junctions made totally by RF magnetron sputtering. *Thin Solid Films* **2015**, *589*, 182–187. [CrossRef]
21. Li, C.C.; Kuo, D.H. Material and technology developments of the totally sputtering-made p/n GaN diodes for cost-effective power electronics. *Mater. Sci. Mater. Electron.* **2014**, *25*, 1942–1948. [CrossRef]

22. Kabra, V.; Aamir, L.; Malik, M. Low cost, p-ZnO/n-Si, rectifying nano heterojunction diode: Fabrication and electrical characterization. *Beilstein J. Nanotechnol.* **2014**, *5*, 2216–2221. [CrossRef] [PubMed]
23. Mohd, M.Z.; Baharin, A.; Hassan, Z.; Abu, H.; Abdullah, M.J. MBE growth of GaN pn-junction photodetector on AlN/Si substrate with Ni/Ag as Ohmic contact. *Superlattices Microstruct.* **2013**, *56*, 35–44. [CrossRef]
24. Cho, S.G.; Nahm, T.U.; Kim, E.K. Deep level states and negative photoconductivity in n-ZnO/p-Si hetero-junction diodes. *Appl. Phys.* **2014**, *14*, 223–226. [CrossRef]
25. Wagner, J.; A. Ramakrishnan, A.; Behr, D.; Maier, M. Composition dependence of the band gap energy of $In_xGa_{1-x}N$ layers on GaN ($x \leq 0.15$) grown by metal-organic chemical vapor deposition. *MRS. Internet. J. Nitri. Semi. Res.* **1999**, *4*, 106–111. [CrossRef]
26. Bedia, F.Z.; Bedia, A.; Benyoucef, B.; Hamzaoui, S. Electrical characterization of n-ZnO/p-Si heterojunction prepared by spray pyrolysis technique. *Phys. Procedia* **2014**, *55*, 61–67. [CrossRef]
27. Baydogan, N.; Karacasu, O.; Cimenoglu, H. Effect of annealing temperature on ZnO:Al/p-Si heterojunctions. *Thin Solid Films* **2012**, *520*, 5790–5796. [CrossRef]
28. Li, J.L.; Schubert, E.F.; Graff, J.W.; Osinsky, A.; Schaff, W.F. Low-resistance ohmic contacts to p-type GaN. *Appl. Phys. Lett.* **2000**, *76*, 2728–2730. [CrossRef]
29. Ponce, A.R.; Olguín, D.; Calderón, H.I. Calculation of the effective masses of II-VI semiconductor compounds. *Superficies y Vacío* **2003**, *16*, 26–28.
30. Suzuki, M.; Uenoyama, T. First-Principles Calculation of Effective Mass Parameters of Gallium Nitride. *Jpn. J. Appl. Phys.* **1995**, *34*, 3442–3446. [CrossRef]
31. Crowell, C.R. The richardson constant for thermionic Emission in schottky barrier diodes. *Solid State Electron.* **1965**, *8*, 395–399. [CrossRef]
32. Kumar, A.; Vinayak, S.; Singh, R. Micro-structural and temperature dependent electrical characterization of Ni/GaN Schottky barrier diodes. *Curr. Appl. Phys.* **2013**, *13*, 1137–1142. [CrossRef]
33. Cheung, S.K.; Cheung, N.W. Extraction of Schottky diode parameters from forward current-voltage characteristics. *Appl. Phys. Lett.* **1986**, *49*, 85–87. [CrossRef]
34. Benamara, Z.; Akkal, B.; Talbi, A.; Gruzza, B. Electrical transport characteristics of Au/n-GaN Schottky diodes. *Mater. Sci. Eng. C* **2006**, *26*, 519–522. [CrossRef]
35. Tuan, T.T.A.; Kuo, D.H. Characteristics of RF reactive sputter-deposited Pt/SiO_2/n-InGaN MOS Schottky diodes. *Mater. Sci. Semicond. Process.* **2015**, *30*, 314–320. [CrossRef]
36. Norde, H. A modified forward I-V plot for Schottky diodes with high series resistance. *J. Appl. Phys.* **1979**, *50*, 5052–5505. [CrossRef]
37. Urgessa, Z.N.; Dobson, S.R.; Talla, K.; Murape, D.M. Optical and electrical characteristics of ZnO/Si heterojunction. *Phys. B* **2014**, *439*, 149–152. [CrossRef]
38. Keskenler, E.F.; Tomakina, M.; Dogan, S. Growth and characterization of Ag/n-ZnO/p-Si/Al heterojunction diode by sol–gel spin technique. *J. Alloy Compd.* **2013**, *550*, 129–132. [CrossRef]
39. Chirakkara, S.; Krupanidhi, S.B. Study of n-ZnO/p-Si (100) thin film heterojunctions by pulsed laser deposition without buffer layer. *Thin Solid Films* **2012**, *520*, 5894–5899. [CrossRef]
40. Shen, Y.; Chen, X.; Yan, X.S. Low-voltage blue light emission from n-ZnO/p-GaN heterojunction formed by RF magnetron sputtering method. *Curr. Appl. Phys.* **2014**, *14*, 345–348. [CrossRef]
41. Hsueh, K.P. Temperature dependent current–voltage characteristics of n-$Mg_xZn_{1-x}O$/p-GaN junction diodes. *Microelectron. Eng.* **2011**, *88*, 1016–1018. [CrossRef]

© 2019 by the authors. Licensee MDPI, Basel, Switzerland. This article is an open access article distributed under the terms and conditions of the Creative Commons Attribution (CC BY) license (http://creativecommons.org/licenses/by/4.0/).

Article

Structural Properties and Oxidation Resistance of ZrN/SiN$_x$, CrN/SiN$_x$ and AlN/SiN$_x$ Multilayered Films Deposited by Magnetron Sputtering Technique

Ihar Saladukhin [1], Gregory Abadias [2,*], Vladimir Uglov [1,3], Sergey Zlotski [1], Arno Janse van Vuuren [4] and Jacques Herman O'Connell [4]

[1] Faculty of Physics, Belarusian State University, 220030 Minsk, Belarus; solodukhin@bsu.by (I.S.); Uglov@bsu.by (V.U.); Zlotski@bsu.by (S.Z.)
[2] Institut Pprime, Université de Poitiers-CNRS-ENSMA, TSA 41123, CEDEX 9, 86073 Poitiers, France
[3] South Ural State University, 454080 Chelyabinsk, Russia
[4] Centre for HRTEM, Nelson Mandela Metropolitan University, Port Elizabeth 6001, South Africa; arnojvv@gmail.com (A.J.v.V.); joconnell@mandela.ac.za (J.H.O.)
* Correspondence: gregory.abadias@univ-poitiers.fr; Tel.: +33-(0)549-496-748

Received: 24 December 2019; Accepted: 4 February 2020; Published: 7 February 2020

Abstract: In the present work, the structure, stress state and phase composition of MeN/SiN$_x$ (Me = Zr, Cr, Al) multilayered films with the thickness of elementary layers in nanoscale range, as well as their stability to high temperature oxidation, were studied. Monolithic (reference) and multilayered films were deposited on Si substrates at the temperatures of 300 °C (ZrN/SiN$_x$ and AlN/SiN$_x$ systems) or 450 °C (CrN/SiN$_x$) by reactive magnetron sputtering. The thickness ratios of MeN to SiN$_x$ were 5 nm/2 nm, 5 nm/5 nm, 5 nm/10 nm and 2 nm/5 nm. Transmission electron microscopy (TEM), X-ray Reflectivity (XRR) and X-ray Diffraction (XRD) testified to the uniform alternation of MeN and SiN$_x$ layers with sharp interlayer boundaries. It was observed that MeN sublayers have a nanocrystalline structure with (001) preferred orientation at 5 nm, but are X-ray amorphous at 2 nm, while SiN$_x$ sublayers are always X-ray amorphous. The stability of the coatings to oxidation was investigated by in situ XRD analysis (at the temperature range of 400–950 °C) along with the methods of wavelength-dispersive X-ray spectroscopy (WDS) and scanning electron microscopy (SEM) after air annealing procedure. Reference ZrN and CrN films started to oxidize at the temperatures of 550 and 700 °C, respectively, while the AlN reference film was thermally stable up to 950 °C. Compared to reference monolithic films, MeN/SiN$_x$ multilayers have an improved oxidation resistance (onset of oxidation is shifted by more than 200 °C), and the performance is enhanced with increasing fraction of SiN$_x$ layer thickness. Overall, CrN/SiN$_x$ and AlN/SiN$_x$ multilayered films are characterized by noticeably higher resistance to oxidation as compared to ZrN/SiN$_x$ multilayers, the best performance being obtained for CrN/SiN$_x$ and AlN/SiN$_x$ with 5 nm/5 nm and 5 nm/10 nm periods, which remain stable at least up to 950 °C.

Keywords: multilayered film; metal nitride; silicon nitride; oxidation

1. Introduction

In accordance with the tendency of industry development, the coatings applied for protection of materials should satisfy more stringent requirements. They have to possess high hardness and wear resistance (e.g., pieces under friction), resistance to high temperature oxidation (e.g., cutting tools) and thermal cyclability (e.g. glass molding dies), stability in corrosive media (e.g., in chemical production units), radiation stability (e.g., materials for nuclear power engineering) and other properties. Currently, physically-vapor deposited transition metal nitride (TMN) coatings based on MeN mononitrides of transition metal (Me = Ti, Zr or Cr) are widely used [1–3]. However, mononitride films often lose their protective role in such severe conditions. For example, TiN and ZrN coatings deposited by reactive magnetron sputtering are intensively oxidized at the temperatures of 500–600 °C [4–7]. This is related to their columnar microstructure and presence of defects (e.g., porosity or micro-cracks), which allows a direct contact between the external atmosphere and substrate and accelerates oxygen incorporation through grain boundary diffusion. Several routes can be employed to enhance oxidation resistance of MeN coatings. Musil et al. [8–10] showed that superior thermal stability, above 1000 °C, was achieved for hard amorphous coatings, based on either ternary Me-Si-N systems with Si content ≥ 20 at.% or quaternary Si-B-C-N system with covalent bonding. These amorphous coatings are generally obtained by co-sputtering method and their improved oxidation resistance is ascribed to the absence of grain boundaries.

An alternate route to nanocomposite/amorphous Me-Si-N single-layer coatings is to deposit sequentially different MeN$_x$ layers, resulting in the formation of periodic multilayers or nanolaminates with improved performance characteristics [11]. By alternation at the nanoscale of dissimilar layers, it is possible to combine the advantages of the different materials properties, and even to get superior properties compared to monolithic films such as improved adhesion, increase in corrosion resistance or change in electrical behavior. The authors of [12] showed that, unlike CrN and AlN films, for which the generation and motion of classical dislocations is responsible for the plastic flow, the observed plastic deformation in CrN/AlN multilayered coatings is mostly governed by grain rotation for the nanocrystals and grain boundary sliding for grains of larger size. Such non-elastic phenomena prevent the deformation or crack formation that allows using CrN/AlN multilayered coatings for ductile steel substrate as well. Another example is the case of AlCrN/TiVN multilayered coatings, where the AlCrN layer inhibited the excessive oxygen diffusion into the multilayered film [13]. Consequently, it was possible to avoid rapid oxidation of vanadium and obtain high wear-resistance for the AlCrN/TiVN coating. Although the TiVN coating had the lowest friction coefficient, the lowest wear rate (1.9 × 10^{-7} mm^3/Nm) was obtained for the AlCrN/TiVN coated sample.

It is possible to classify three main types of multilayers based on the type of structure of the elementary layers: (i) nanocrystalline/nanocrystalline layers; (ii) nanocrystalline/amorphous layers; and (iii) amorphous/amorphous layers.

Multilayers based on TMN usually belong to the first category [12–17]. The comparison of hardness and other mechanical characteristics of multilayered coatings made of alternate stacking of various TMN is given in [15]. It is pointed that the TiN/VN multilayered coatings are one of the most successful examples of superhard materials. The maximum hardness value of these coatings reached 56 GPa (at bilayer thickness about 5.2 nm). The use of a combination of nanocrystalline/nanocrystalline layers is rather effective for achievement of high resistance of the coatings to high temperature oxidation [16,17]. High thermal stability of AlCrN/TiSiN coating is noted, for which the excellent oxidation resistance was obtained with no pore and delamination up to 900 °C [17]. In this case, the redistribution of elements between the layers takes place when forming the oxide layer, namely mixed (Al,Ti)-oxide scale outside layer and a dense (Al,Cr)$_2$O$_3$ inner layer.

The second group of multilayered coatings, consisting of sequentially alternating layers of nanocrystalline and amorphous material, is perspective due to the possibility of combining of the different properties. For example, the hardness and wear resistance of nanocrystalline layer are supplemented with plasticity and chemical inertness of the amorphous layer. MeN/SiN$_x$ coatings are among the most appealing nanocrystalline/amorphous layered systems [18–21]. The characteristic feature of these films is the immiscibility of their constitutive layers at the origin of their good thermal stability and properties [20,22,23]. The enhanced stability of MeN/SiN$_x$ films to oxidation is also related to the fact that the columnar growth of MeN layer is suppressed by the presence of the SiN$_x$ amorphous layer [21]. This prevents from oxygen penetration deep into the film along the grain boundaries and pores concomitant to the formation of the columnar structure.

A significant increase in hardness of the MeN/SiN$_x$ multilayered films has been reported when decreasing the thickness of SiN$_x$ layer below ~0.8 nm [23–25]. In this case, the epitaxial growth of silicon nitride at the interface of cubic MeN layer results in SiN_x layer crystallization. Such an effect is more pronounced for TiN/SiN_x and ZrN/SiN_x systems [18,24–26], and is less marked for the AlN/SiN_x system because the interfaces between AlN and epitaxial Si_3N_4 only slightly affect the propagation of dislocations. As the result, only a minor hardness enhancement is expected in AlN/SiN_x multilayers [27]. However, it should be noted that an increase in microhardness of MeN/SiN_x multilayered coatings at the thicknesses of SiN_x elementary layer of 1 nm and less is accompanied by a decrease in film oxidation resistance [6]. This indicates that, by controlling the ratio of individual layers, it is possible to ensure one or another required property.

Among the coatings of the third group (amorphous/amorphous layers), the coatings with alternating layers of so-called metallic glasses are of more interest for practical applications. ZrCu/ZrCuNiAlSi films are an example of radiation-resistant multilayered systems for applications in the area of materials design for nuclear power engineering [28].

In our prior work on the ZrN/SiN_x system, we showed that the coatings which belong to the second group are quite promising for use as resistant coatings under high-temperature oxidation up to 860–950 °C [6]. It should be noted that, for the CrN/SiN_x system and, especially, for the AlN/SiN_x system, the chemical inertness of SiN_x amorphous layer is supplemented by the formation of the passivating layers of metal oxide in the surface layer of the MeN/SiN_x multilayered film [17,29]. However, it is difficult to compare the oxidation behavior of the different MeN/SiN_x multilayered systems because of the different physical vapor deposition processes used for their fabrication, as well as variations in the bilayer thickness range studied. In the present work, we performed a systematic and comparative study of the structure and stability to high-temperature oxidation of ZrN/SiN_x, CrN/SiN_x and AlN/SiN_x films sputter-deposited in the same chamber, with special emphasis laid on the influence of the thickness ratio of the elementary layers. As shown previously for the ZrN/SiN_x system, the ratio of thicknesses of individual layers plays a key role in their oxidation resistance [6]. The properties of the multilayers are also discussed and compared to those of reference monolithic films.

2. Materials and Methods

Reference ZrN, CrN, AlN and Si_3N_4 monolithic films as well as ZrN/SiN_x, CrN/SiN_x and AlN/SiN_x multilayered films were grown by reactive magnetron sputter-deposition in a high vacuum chamber (base pressure $< 10^{-5}$ Pa) equipped with confocal targets configuration and a cryogenic pump (max. 500 L/s) [30]. All films were deposited on Si substrates covered with 10 nm thick thermally grown SiO_2 layer (to prevent the interdiffusion of the coating and substrate components). A constant bias voltage of −60 V was applied to the substrate during deposition. The substrate was rotated at 15 rpm throughout the deposition to ensure an equal deposition rate across the substrate surface.

Water-cooled, 7.62-cm-diameter Zr (99.2% purity), Cr (99.95% purity), Al (99.9995% purity) and Si_3N_4 (99.99% purity) targets, located at 18 cm from the substrate holder, were used under Ar + N_2 plasma discharges at constant power mode. The Zr, Cr and Al targets were operated in magnetically unbalanced configuration using a DC power supply, while for the Si_3N_4 target a RF power supply was used in balanced mode. The total working pressure varied from 0.22–0.29 Pa depending on the material system, as measured using a Baratron® capacitance gauge (Andover, MA, USA). The Ar/N_2 flow ratio was optimized to obtain stoichiometric nitrogen content in the films based on earlier results [31,32] (see Table 1). The N_2 partial pressure was measured during deposition using a MKS Microvision mass spectrometer (Munich, Germany).

The deposition conditions were not identical for all MeN layers. They varied (N_2 partial pressure and substrate temperature) to achieve the desired phase and optimal crystallinity for each MeN layer. Reference ZrN, AlN and Si_3N_4 films as well as ZrN/SiN_x and AlN/SiN_x multilayered films were deposited at 300 °C, while reference CrN film and CrN/SiN_x multilayered films were formed at the substrate temperature of 450 °C. This temperature was chosen to obtain single-phase cubic CrN phase without formation of hexagonal Cr_2N phase. The periodic growth of the multilayered systems was monitored by computer-controlled pneumatic shutters located at 2 cm in front of each target. Details of the multilayered films growth procedure are given in [23].

Table 1. Process parameters and chemical analysis for ZrN, CrN, AlN and Si$_3$N$_4$ reference films.

Film	T$_{dep.}$ (°C)	Me Power (W)	Si$_3$N$_4$ Power (W)	Ar/N$_2$ Flow	Total Pressure(Pa)	Partial Pressure P$_{N_2}$ (Pa)	Growth Rate(nm/min)	Film Thick-ness (nm)	Me (at. %)	Si (at. %)	N (at. %)
ZrN	300	300	–	10/0.5	0.22	4.6 × 10^{-3}	12.5	249	46.6	–	53.4
CrN	450	200	–	25/20	0.29	6.3 × 10^{-2}	5.4	259	54.6	–	45.4
AlN	300	300	–	24/6.5	0.22	2.4 × 10^{-2}	5.9	293	43.6	–	56.4
Si$_3$N$_4$	300	–	176	24/5.1	0.22	2.2 × 10^{-2}	2.1	287	–	43.3	56.7

Note. For ZrN/SiN$_x$, CrN/SiN$_x$ and AlN/SiN$_x$ films, the same deposition regimes as for ZrN, CrN, AlN and Si$_3$N$_4$ reference films were used.

The deposition conditions are summarized in Table 1 for the reference monolithic films, and the same conditions were used for elementary layers of the MeN/SiN$_x$ (Me = Zr, Cr, Al) multilayers, except the deposition time. The deposition rates were 12.5, 5.9 and 5.4 nm/min for ZrN, AlN and CrN monolithic films, respectively. The use of a RF discharge to deposit the Si$_3$N$_4$ reference film resulted in a slower deposition rate of 2.1 nm/min. For mononitrides, the deposition time was adjusted between 22 min and 2.3 h to get similar nominal film thickness of ~300 nm. For MeN/SiN$_x$ multilayers, the deposition time of constitutive sublayers was adjusted accordingly to get the desired modulation period, but the total film thickness was similar (~300 nm). For example, for MeN (5 nm)/SiN$_x$ (5 nm), the deposition times were 23 s/141 s, 50 s/141 s and 50 s/141 s for MeN=ZrN, AlN and CrN, respectively. The nominal ratios of thickness of MeN layer to SiN$_x$ layer were 5 nm/2 nm, 5 nm/5 nm, 5 nm/10 nm and 2 nm/5 nm (see Table 2). For multilayers, the deposition process always started with the MeN sublayer being deposited first.

Table 2. Best-fit parameters as determined from simulations of XRR experimental data of MeN/SiN$_x$ (Me = Zr, Cr, Al) multilayers: elementary layer thickness (h_{MeN} and h_{SiNx}), mass density (ρ_{MeN} and ρ_{SiNx}), interface roughness (w_{MeN} and w_{SiNx}), MeN layer fraction (f_{MeN}) and total film thickness. The first row indicates nominal thicknesses ratio for MeN/SiN$_x$ multilayers.

Multi-Layers	MeN (Me = Zr, Cr, Al) Sublayer			SiN$_x$ Sublayer			f_{MeN}	Number of Bilayers	Total Thickness (nm)
	h_{MeN} (nm)	ρ_{MeN} (g cm^{-3})	w_{MeN} (nm)	h_{SiNx} (nm)	ρ_{SiNx} (g cm^{-3})	w_{SiNx} (nm)			
				ZrN/SiN$_x$					
5 nm/5 nm	3.7	7.3	0.4	5.3	3.1	0.6	0.41	29	261
5 nm/10 nm	4.9	7.4	0.7	9.9	2.9	0.5	0.33	20	296
2 nm/5 nm	1.6	7.4	0.6	5.0	3.0	0.6	0.24	43	284
				CrN/SiN$_x$					
5 nm/5 nm	4.5	6.2	0.6	4.9	3.0	0.3	0.48	30	282
5 nm/10 nm	4.4	6.4	0.4	9.7	3.0	0.3	0.31	20	282
2 nm/5 nm	1.6	6.2	0.3	5.1	3.0	0.3	0.24	43	288
				AlN/SiN$_x$					
5 nm/5 nm	4.3	3.2	0.2	5.6	3.0	0.8	0.43	30	297
5 nm/10 nm	3.8	3.3	<0.1	11.3	3.0	0.4	0.25	20	302
2 nm/5 nm	1.6	3.2	0.2	5.6	3.0	0.8	0.22	43	310

The evolution of intrinsic stress developed during growth was monitored in situ and in real-time using the wafer curvature technique [33]. A multiple-beam optical stress sensor (MOSS) designed by kSpace Associates (kSA, Dexter, MI, USA) was implemented in the deposition chamber. The measurements were performed using 150 ± 2 µm thick Si substrate covered with native oxide, under stationary mode.

High resolution transmission electron microscopy (HRTEM, JEOL JEM ARM200F, Tokyo, Japan) analysis was carried out for direct information about the film structure and state of interlayer boundaries. Cross-sectional TEM specimens were prepared using a FEI Helios Nanolab 650 focused ion beam (FIB) (Brno, Czech Republic). More details on the sample preparation can be found in [6]. All specimens were analyzed using a JEOL JEM 2100 LaB$_6$ transmission electron microscope (Tokyo, Japan) operating at 200 kV.

The characterization of the multilayer stacking was carried out using low-angle X-ray Reflectivity (XRR). A fitting procedure, based on the optical formalism of Parratt [34], was used to extract the relevant quantities (individual layer thickness, mass density and interface roughness). The fraction of MeN (Me = Zr, Cr, Al) layer is defined as $f_{MeN} = \frac{h_{MeN}}{h_{MeN}+h_{SiN_x}}$, where h_{MeN} and h_{SiN_x} are the MeN and SiN$_x$ layer thicknesses, respectively.

The elemental composition of films in their as-deposited (for the reference films only) and air-annealed states was determined using elemental probe microanalysis. A wavelength dispersive spectrometer (WDS) unit from Oxford Instruments (High Wycombe, UK) attached to a JEOL 7001 TTLS scanning electron microscope (Tokyo, Japan) operated at 10 kV and 10 nA was used for the quantification with a precision better than 1 at.%. The same microscope was applied for obtaining top-view SEM micrographs of the films after air annealing at 950 °C.

X-ray Diffraction (XRD) analysis was employed for structural identification using a Bruker D8 AXS X-ray diffractometer (Karlsruhe, Germany) operating in Bragg–Brentano configuration and equipped with Cu$_{K\alpha}$ wavelength (0.15418 nm) and LynxEye detector.

Isothermal air annealing was performed at different sequential temperatures from 400 to 950 °C. The oxidation process was investigated using in situ XRD experiments. The samples were placed on a resistive heating stage designed by Anton Paar (Graz, Austria) implemented on the Bruker D8 diffractometer, consisting in an AlN sample holder and a hemispheric graphite dome subjected to air blowing. Total scan time during isothermal annealing was 50–60 min.

3. Results and Discussion

3.1. Structure and Phase Composition of As-Deposited MeN/SiN$_x$ (Me = Zr, Cr, Al) Multilayered Films

As described in Section 2, the ZrN/SiN$_x$, CrN/SiN$_x$ and AlN/SiN$_x$ multilayered films with different thickness ratio of MeN (Me = Zr, Cr, Al) and SiN$_x$ elementary layers were formed. This corresponds to different MeN fractions, f_{MeN}. The elemental composition of the reference monolithic films, obtained from WDS, is given in Table 1. MeN$_x$ films are slightly off-stoichiometric: overstoichiometric ZrN$_{1.15}$ and AlN$_{1.30}$ films and substoichiometric CrN$_{0.83}$ films were obtained. However, the SiN$_x$ film has the Si$_3$N$_4$ stoichiometry.

Figure 1 shows the HRTEM images of MeN/SiN$_x$ (Me = Zr, Cr, Al) multilayered films with the ratio of elementary layer thicknesses equal to 5 nm/5 nm. It reveals the uniform alternation of MeN and SiN$_x$ elementary layers, and the formation of sharp and smooth interfaces between the layers points to the absence of intermixing of the layer components. However, the interlayer boundary seems more diffuse for the CrN/SiN$_x$ multilayer, possibly due to the growth at higher substrate temperature. While the MeN$_x$ layers appear crystalline (see, e.g., lattice fringes in AlN layer of Figure 1c), the SiN$_x$ layers were found to be amorphous. Note also a change in electronic contrast from ZrN/SiN$_x$ to AlN/SiN$_x$ multilayers, due to reduction in mass density difference between MeN and SiN$_x$ layers. The values of mass densities for the elementary layers are presented in Table 2, as extracted from the fit of XRR data (see below). In particular, the contrast between AlN and SiN$_x$ layers is very similar, as also observed by other authors [27].

Figure 2 shows the low-angle XRR scans together with the simulated curves obtained for MeN/SiN$_x$ (Me = Zr, Cr, Al) multilayered films with the nominal MeN thickness h_{MeN} = 5 nm and SiN$_x$ thickness h_{SiN_x} = 5 nm. For all multilayered systems, relatively sharp superlattice reflections are observed up to high 2θ angles (up to seventh order or more). This testifies the formation of highly periodic multilayer structures and gives evidence of the presence of relatively smooth interface boundaries between the layers. Similar results were obtained for CrN/Si$_3$N$_4$ multilayered films in [20], where the effectiveness of Si$_3$N$_4$ layer for stabilization of periodic structure was mentioned. Note that Kiessig fringes are also distinguishable between main superlattice reflections in XRR scans (however, not visible in the scale displayed in Figure 2).

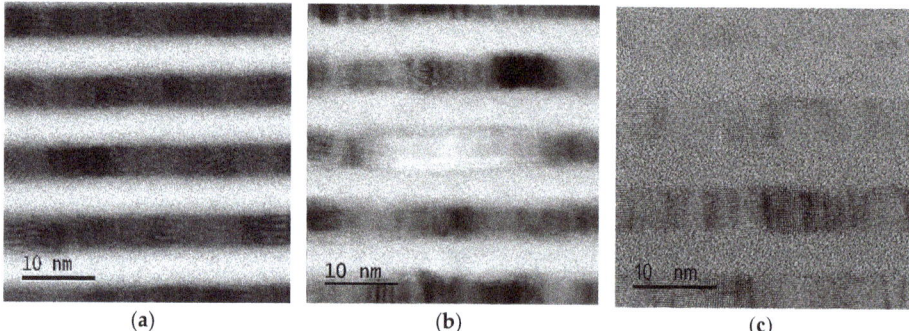

Figure 1. Cross-sectional (bright field) HRTEM images of multilayered films with the ratio of thicknesses of MeN (darker contrast) and SiN_x (brighter contrast) elementary layers equal to 5 nm/5 nm: ZrN/SiN_x (**a**); CrN/SiN_x (**b**); and AlN/SiN_x (**c**).

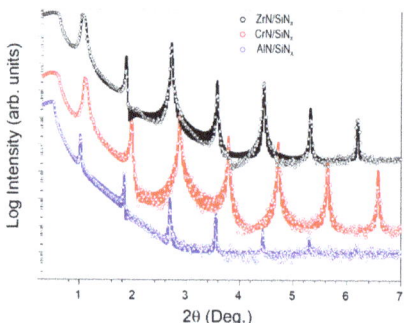

Figure 2. XRR scans of ZrN/SiN_x, CrN/SiN_x and AlN/SiN_x multilayered films with the ratio of thicknesses of MeN and SiN_x elementary layers equal to 5 nm/5 nm. Solid lines correspond to the best-fit simulations to experimental data using optical model of Parratt.

An optical simulation model was used to fit XRR data and get information on layer density, layer thickness and interface roughness of each sublayer for ZrN/SiN_x, CrN/SiN_x and AlN/SiN_x multilayers. Values are reported in Table 2 for the different systems and bilayer periods investigated. The obtained results show that the boundaries between the layers for all multilayered systems are quite smooth (the interface roughness w_{MeN} does not exceed 0.6 nm), which is in good agreement with the HRTEM observations presented in Figure 1. All MeN_x sublayers were found to be dense, with mass density of $\rho_{ZrN} = 7.4 \pm 0.1$ g·cm^{-3}, $\rho_{CrN} = 6.3 \pm 0.1$ g·cm^{-3} and $\rho_{AlN} = 3.2 \pm 0.1$ g·cm^{-3}. These values are very close to the values for bulk reference powders, being $\rho_{ZrN} = 7.29$ g·cm^{-3}, $\rho_{CrN} = 6.18$ g·cm^{-3} and $\rho_{AlN} = 3.26$ g·cm^{-3}. The value ρ_{SiNx} was found to be equal to 3.0 g· $\pm$ 0.1 cm^{-3}, also in good agreement with the bulk value of α-Si_3N_4 crystalline phase (3.18 g·cm^{-3}).

One can note from the results in Table 2 that for all multilayered films the thickness of MeN (Me = Zr, Cr, Al) layer is less than the nominal thickness. For example, for the case of nominal ratio of MeN and SiN_x thicknesses 2 nm/5 nm, the real thickness of MeN layer is 1.6 nm, i.e. 20% less. This is related to the poisoning of the metal target during the first seconds of the reactive sputter-deposition process. Consequently, the initial deposition rate is lower than the average value calculated from the thicker reference MeN films. More information on this poisoning effect can be found in [23]. At the same time, the thickness of SiN$_x$ layer is close to nominal one for all multilayered systems (Table 2).

Figure 3 shows the substrate curvature change measured by MOSS during growth of ZrN/SiN$_x$, CrN/SiN$_x$ and AlN/SiN$_x$ multilayered films with nominal thickness ratio 10 nm/5 nm. One can see from the slope of F/w vs. layer thickness that both sublayers develop compressive stress. However, the magnitude of the compressive stress is much larger for MeN layers than for SiN$_x$ layers. Relatively large values of −7.2 and −5.7 GPa were obtained for the incremental stress during deposition of AlN and ZrN layers, respectively, while for CrN layers the incremental stress is −2.7 GPa. Comparatively, the incremental stress of SiN$_x$ layer is about −1.2 GPa. One can also notice from the results in Figure 3 that the stress state is reproducible from one bilayer to another, i.e., there is no influence of the underneath layers on the cumulative stress build-up. The formation of compressive stress in sputter-deposited TMN layers is due to energetic bombardment during growth, which creates point defects in the crystal lattice and densify the grain boundaries [23,35–37]. At low deposited energy, sputter-deposited MeN$_x$ films develop a columnar, underdense microstructure, often accompanied by the development of tensile stress [35,37,38]. This is obviously not the case here for these nanoscale layer thicknesses, because the presence of amorphous SiN$_x$ layers interrupts the columnar growth of MeN layers. If the energy delivered to growing film is large, a dense and sometimes featureless microstructure is formed together with compressive stress. In our deposition conditions, the main contribution to the deposited energy stems from energetic neutrals (sputtered atoms and backscattered Ar) due to low fraction of bombarding ions (a few percent) reaching the substrate. The lower compressive stress values for CrN layers is likely due to the higher substrate temperature (450 °C vs. 300 °C), which favors defect annihilation processes. Amorphous layers are more tolerant to defect incorporation compared to their crystalline counterpart, which explains the lower stress values obtained for SiN$_x$.

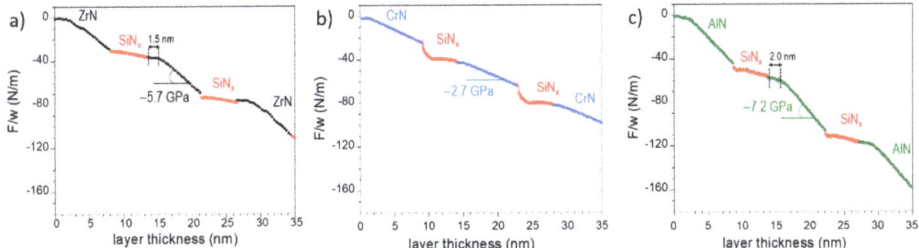

Figure 3. In situ stress evolution during growth of multilayered films: ZrN/SiN$_x$ (**a**); CrN/SiN$_x$ (**b**); and AlN/SiN$_x$ (**c**). The data were obtained from MOSS and are displayed as the variation of the force per unit width (F/w) vs. layer thickness. The slope of the curves gives the information on the incremental stress, ranging from −7.2 GPa (AlN) to −2.7 GPa (CrN).

A closer inspection to the stress curves of Figure 3 reveals some interesting features. The development of compressive stress starts after ~1.5 nm and 2.0 nm for ZrN and AlN layers, respectively. Below these values, the stress is relatively small (about −1 GPa) and could reflect the initial formation of an amorphous layer. XRD characterizations on the MeN/SiN$_x$ multilayers with 2 nm/5 nm thickness ratio support this scenario. Finally, it can be noticed in Figure 3b that the stress development during growth of SiN$_x$ on CrN is peculiar, as a steady-state stress is only reached after ~2–3 nm. The larger compressive stress that develops in the very beginning is contributed to interface stress and possible interlayer formation, as also noticed from cross-section HRTEM view in Figure 1b.

In Figures 4–6 (see black lines for as-deposited state), the XRD patterns of reference monolithic ZrN (Figure 4a), CrN (Figure 5a) and AlN (Figure 6a) films are represented in comparison with the XRD patterns of MeN/SiN$_x$ (Me = Zr, Cr, Al) multilayered films (Figure 4b–d, Figures 5b–d and 6b–d). The angular range 25–58° covers the main 111 and 200 Bragg reflections for cubic c-ZrN (JCPDS card

No. 35-0753), c-CrN (JCPDS card No. 76-2494) as well as 100, 002, 101 and 102 reflections for hexagonal h-AlN with wurtzite structure (JCPDS card No. 25-1133).

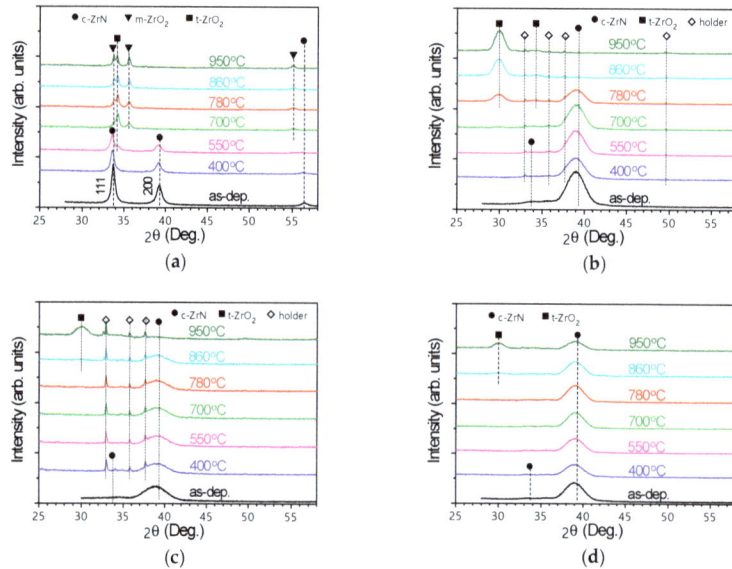

Figure 4. Evolution of XRD patterns under air annealing for ZrN reference film (**a**); and ZrN/SiN$_x$ multilayered films with different thickness of ZrN and SiN$_x$ elementary layers: (**b**) 5 nm/2 nm; (**c**) 5 nm/5 nm; and (**d**) 5 nm/10 nm.

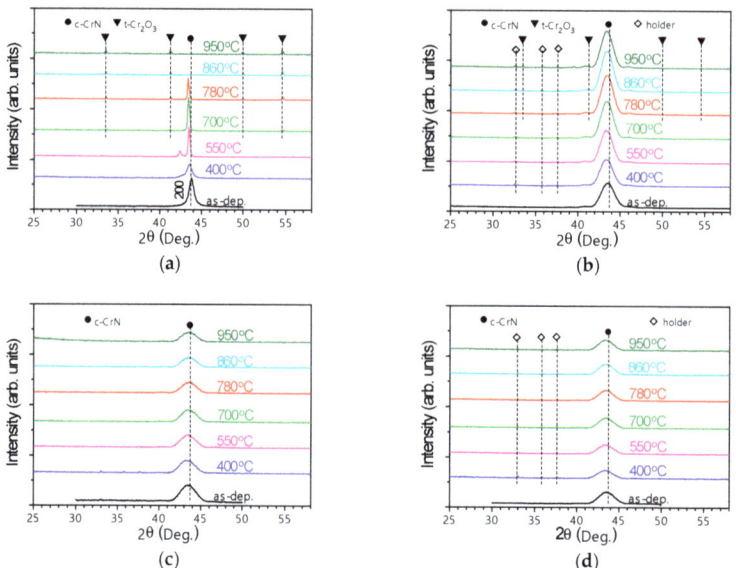

Figure 5. Evolution of XRD patterns under air annealing for CrN reference film (**a**); and CrN/SiN$_x$ multilayered films with different thickness of CrN and SiN$_x$ elementary layers: (**b**) 5 nm/2 nm; (**c**) 5 nm/5 nm; and (**d**) 5 nm/10 nm.

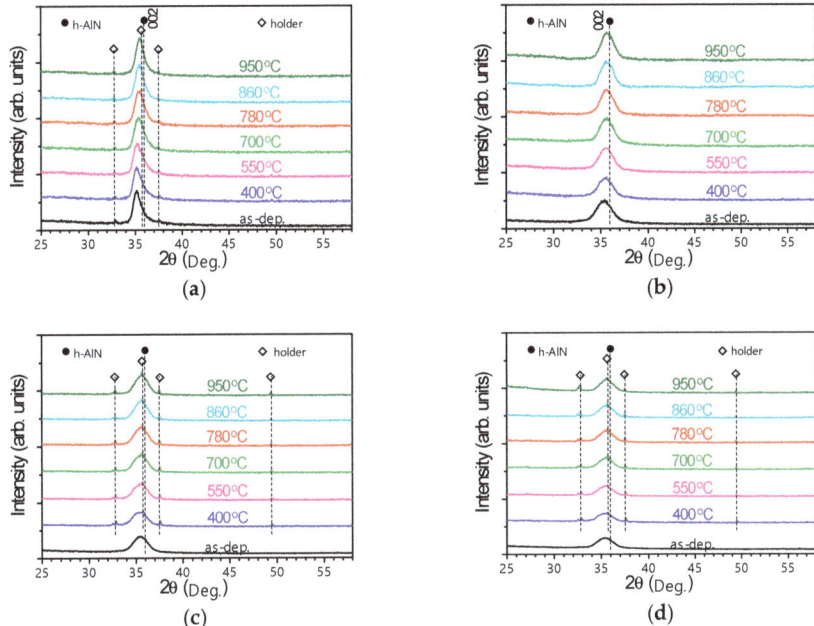

Figure 6. Evolution of XRD patterns under air annealing for AlN reference film (**a**); and AlN/SiN$_x$ multilayered films with different thickness of AlN and SiN$_x$ elementary layers: (**b**) 5 nm/2 nm; (**c**) 5 nm/5 nm; and (**d**) 5 nm/10 nm.

If for the reference c-ZrN film (Figure 4a) both (111) and (200) preferred orientations take place, in the case of reference c-CrN and h-AlN films, the preferred orientation along the [200] direction (c-CrN) or in the [002] direction (h-AlN) is only observed (Figures 5a and 6a). Competitive columnar growth between (111)- and (200)-oriented ZrN crystallites during sputter-deposition of ZrN monolithic films was discussed previously [23,35]. The (200) preferred orientation is characteristic for CrN films [39,40], as well as (002) preferred orientation is typical for h-AlN films [41–43]. Torino et al. [42] reported a transition from (101) to (002) preferred orientation for AlN films with decreasing (Ar + N$_2$) working pressure in the chamber. The presence of (002) orientation means that the AlN crystallites are highly oriented with the c-axis perpendicular to the substrate surface [43,44].

It was observed that the peak position of reference ZrN, CrN and AlN monolithic films is shifted to lower 2θ angles as compared to the position for bulk materials. This is related to the presence of in-plane compressive stresses, resulting in an expansion of the out-of-plane lattice parameter. For the AlN reference film, an asymmetry of the 002 peak towards higher 2θ angles was noticed, which can be due to a higher defect concentration. The reference Si$_3$N$_4$ film was found to be X-ray amorphous, i.e., there are no reflections in the investigated angular range (see [23]).

If we now analyze the XRD patterns of MeN/SiN$_x$ (Me = Zr, Cr, Al) multilayers in their as-deposited state, the following observations can be made in comparison with the reference MeN films: (i) a transition from (111) preferred orientation, which predominates for reference ZrN film (Figure 4a), to (200) orientation for ZrN/SiN$_x$ multilayered films (Figure 4b–d) and the retaining of (200) or (002) preferred orientation for CrN/SiN$_x$ (Figure 5b–d) or AlN/SiN$_x$ (Figure 6b–d) multilayered films as in the case of corresponding mononitrides CrN (Figure 5a) or AlN (Figure 6a); (ii) the broadening of the MeN XRD lines and decrease in their intensity that can be caused by the decrease in MeN crystallites size; and (iii) the amorphization of the MeN/SiN$_x$ (Me = Zr, Cr, Al) multilayered films when the MeN layer thickness decreases down to 2 nm (XRD patterns of the ZrN/SiN$_x$ multilayer with 2 nm/5 nm thickness ratio can be found in our previous works [6,23]).

In the case of ZrN/SiN$_x$ multilayered films, the amorphous SiN$_x$ layer hinders the columnar growth of ZrN crystallites and favors the (200) preferred orientation [23]. The transition to (200) preferred orientation and broadening of ZrN peak with decreasing f_{MeN} was also observed by Dong et al. [25]. CrN/SiN$_x$ multilayered films were also characterized by (200) preferred orientation [20]. The present results allow concluding that the insertion of amorphous SiN$_x$ layers influence the structure of growing MeN layers.

As for the reference MeN$_x$ films, the position of ZrN and CrN 200 lines, and AlN 002 line is shifted to lower angles compared to position for bulk materials, in good agreement with an in-plane compressive stress state, as revealed from substrate curvature measurements (see Figure 3).

3.2. Evolution of Phase Composition of MeN/SiN$_x$ (Me = Zr, Cr, Al) Multilayered Films during Air Annealing

The evolution of XRD patterns for reference (monolithic) single-layer MeN and for multilayered films of different composition with the increase of air-annealing temperature from of 400 to 950 °C is shown in Figures 4–6. When considering the reference ZrN (Figure 4a), CrN (Figure 5a) and AlN (Figure 6a) films, it is worth noting the following. Oxidation of ZrN starts already at the temperature of 550 °C (t-ZrO$_2$ phase is registered), oxidation of CrN at the temperature of 700 °C (diffraction reflections of t-Cr$_2$O$_3$ phase appear), while the AlN film remains unaltered up to 950 °C. This shows the higher thermal stability of AlN. The peaks of ZrN and CrN phases disappear completely when reaching the temperature of 700 and 860 °C, respectively. It should be noted that the shoulder to the left from 200 c-CrN peak is detected for CrN film at the temperature of 400 °C, and afterwards the weak 111 reflection of h-Cr$_2$N phase (JCPDS card No. 35-0803) at 2θ ≈ 42.6° is registered at 550 °C (Figure 5a). Previous studies have also reported the transformation from c-CrN to h-Cr$_2$N phase during vacuum or air annealing [45,46] due to depletion in nitrogen. Both phases decompose with the formation of chromium oxides at the subsequent temperature rise.

In the case of ZrN/SiN$_x$ multilayered films (Figure 4b–d), the evolution of XRD patterns during air annealing depends on the f_{MeN} fraction. For ZrN/SiN$_x$ (5 nm/2 nm), for which f_{MeN} > 0.50, the crystallization of t-ZrO$_2$ oxide phase occurs at the temperatures of 700–780 °C (Figure 4b). When reaching the temperature of 860 °C, the 200 ZrN peak disappears, indicating the decomposition of the nitride phase and formation of zirconium oxides. For ZrN/SiN$_x$ (5 nm/5 nm) and ZrN/SiN$_x$ (5 nm/10 nm) films, for which f_{MeN} < 0.50 (Table 2), the crystallization of t-ZrO$_2$ takes place at the higher temperatures (860–950 °C). In the case of ZrN/SiN$_x$ (5 nm/10 nm), the ZrN peak remains until 950 °C (Figure 4d). Therefore, it can be concluded that oxidation resistance of ZrN/SiN$_x$ multilayers increases with increasing SiN$_x$ layer thickness from 2 to 10 nm at the constant thickness of 5 nm for ZrN layer. A similar trend of thermal stability enhancement with decreasing f_{MeN} fraction was reported earlier [6]. Note that the ZrN/SiN$_x$ (2 nm/5 nm) amorphous multilayer film was also stable in the temperature range of 400–950 °C, with no oxide phases detected (see Figure 13d in [6]). The value of f_{MeN} for this multilayer is similar to that for ZrN/SiN$_x$ (5 nm/10 nm) multilayer but the interface density is more than double (Table 2). This suggests that an increase in interface density also promotes the oxidation resistance of ZrN/SiN$_x$ multilayered films.

Contrarily to the reference CrN film (Figure 5a), CrN/SiN$_x$ multilayers were found to be thermally stable up to 950 °C. No crystalline oxide phases are detected up to temperature of 950 °C. For the CrN/SiN$_x$ (5 nm/2 nm) multilayered films (Figure 5b), the intensity of 200 CrN peak even slightly increases with temperature, which is most likely connected to some improvement of crystalline quality of this film. The CrN/SiN$_x$ (2 nm/5 nm) film, which was amorphous in its as-deposited state, remains amorphous after air annealing (not shown).

For the third multilayered system, i.e. for AlN/SiN$_x$ films, the emergence of oxide phases during air annealing is not registered for all studied thickness ratios of elementary layers, namely, 5 nm/2 nm, 5 nm/5 nm, 5 nm/10 nm and 2 nm/5 nm (Figure 6b–d). Intensity of 002 AlN peak rises slightly with the increase in annealing temperature for all samples except for AlN/SiN$_x$ (2 nm/5 nm) film, which remains X-ray amorphous.

To get more insights on phase stability, we plot in Figure 7 the evolution of the out-of-plane lattice parameter of MeN layers with annealing temperature. Lattice parameter a of c-ZrN and c-CrN phases was calculated using angular position of 200 peak, and lattice parameter c of h-AlN phase using 002 peak. The results are shown for the reference ZrN, CrN and AlN films, as well as for MeN/SiN$_x$ (Me = Zr, Cr, Al) multilayers with 5 nm/5 nm and 5 nm/10 nm ratios. For AlN sub-layers, as well as AlN monolithic film, a substantial decrease of the lattice parameter was found, up to 0.7% relative reduction for the reference AlN film. This is contributed to the relaxation of compressive stress with increasing annealing temperature, which decreases propensity of film delamination and/or bucking and is therefore beneficial to its thermal stability. An opposite behavior was found for ZrN and CrN reference films, which could only be explained by decomposition of the MeN phase due to nitrogen release. For ZrN/SiN$_x$ and CrN/SiN$_x$ multilayers, one observes a competition between nitrogen loss and relaxation of compressive stress, depending on the annealing temperature.

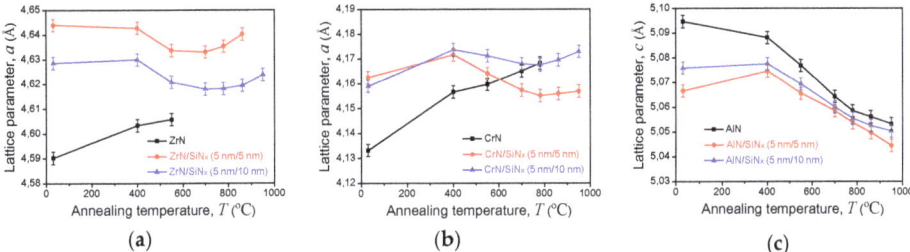

Figure 7. Dependences of the lattice parameter of MeN phase on the annealing temperature for multilayered films with comparison with the same dependences for reference monolithic films: ZrN/SiN$_x$ (**a**); CrN/SiN$_x$ (**b**); and AlN/SiN$_x$ (**c**).

3.3. Elemental Composition and Surface State of MeN/SiN$_x$ (Me = Zr, Cr, Al) Multilayered Films after Air Annealing Procedure

The composition of the reference ZrN, CrN, AlN and Si$_3$N$_4$ films as well as MeN/SiN$_x$ (Me = Zr, Cr, Al) multilayered films after air annealing up to 950 °C was measured by WDS. Concerning the multilayered films, it should be noted that by means of WDS the integral content of the elements (corresponding to the whole thickness of the coating) was determined. Consequently, the obtained data allowed carrying out the relative comparison of oxidation resistance of the multilayered films of three systems—ZrN/SiN$_x$, CrN/SiN$_x$ and AlN/SiN$_x$—with the same ratios of elementary layer thicknesses. The oxygen and nitrogen content in the film composition gives information on film stability after air-annealing. As the oxygen incorporates and substitutes the nitrogen atoms in MeN lattice during high-temperature annealing, the increase in oxygen content in the coating composition is always accompanied by a decrease in nitrogen content. The variation in oxygen content for the reference and multilayered films is displayed in Figure 8.

If the ZrN and CrN films are fully oxidized (the oxygen content is close to composition of ZrO$_2$ and Cr$_2$O$_3$ oxides, respectively), the AlN and Si$_3$N$_4$ films are characterized by a considerably lower degree of oxidation (around 30 at.%, see Figure 8). Overall, the MeN/SiN$_x$ multilayers show better oxidation resistance than their mononitride counterparts, in agreement with XRD analysis. For the ZrN/SiN$_x$ system, the tendency of oxidation resistance enhancement with decreasing f_{MeN} was clearly observed. In the case of CrN/SiN$_x$ and AlN/SiN$_x$ systems, the same tendency is revealed but it is much less pronounced. It is noteworthy that for CrN/SiN$_x$ and AlN/SiN$_x$ films the decrease in MeN layer thickness down to 2 nm leads to a certain deterioration of oxidation resistance.

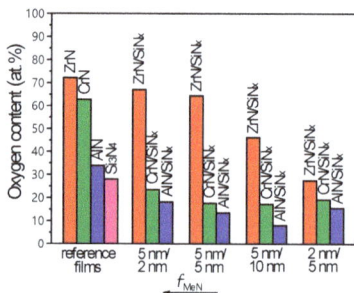

Figure 8. The oxygen content for the reference ZrN, CrN, AlN, and Si$_3$N$_4$ films as well as ZrN/SiN$_x$, CrN/SiN$_x$ and AlN/SiN$_x$ multilayered films with different ratios of the thicknesses of MeN and SiN$_x$ elementary layers after annealing in air at 950 °C.

The results reported in Figure 8 allow concluding that CrN/SiN$_x$ and AlN/SiN$_x$ multilayered films are significantly more stable under conditions of high-temperature annealing as compared to ZrN/SiN$_x$ films. These results agree with XRD data discussed in the previous section. The smallest content of oxygen (8 at. %) after annealing in air was registered for AlN/SiN$_x$ (5 nm/10 nm) film.

To clarify the mechanisms responsible for film oxidation, the analysis of the surface state was performed by SEM after high-temperature (950 °C) annealing. SEM observations reveal that the high degree of the coating damage, namely the emergence of corrosion sites, the swelling and flacking of the film, is inherent to the reference monolithic films, except for AlN surface, which remains quite uniform after annealing (see Figure 9).

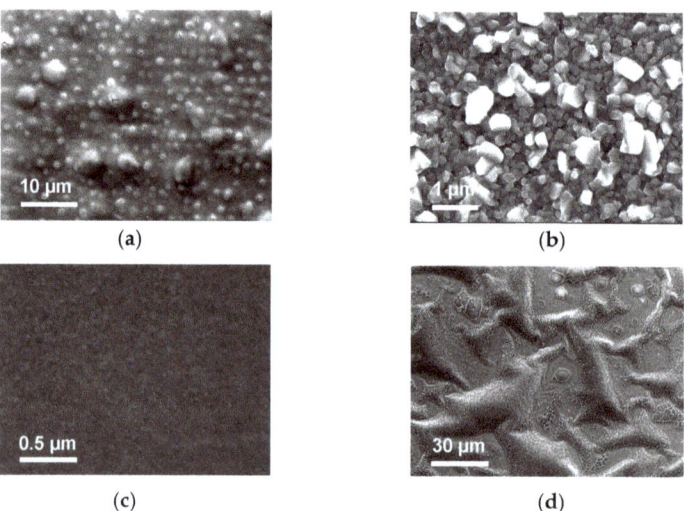

Figure 9. Top-view SEM micrographs of reference films after air annealing at 950 °C: ZrN (**a**); CrN (**b**); AlN (**c**); and Si$_3$N$_4$ (**d**). Note the different scale bar for the different images.

The MeN/SiN$_x$ (Me = Zr, Cr, Al) multilayered films are appreciably less subjected to surface damage after air annealing. Besides, the analysis of the surface topography of CrN/SiN$_x$ and AlN/SiN$_x$ multilayered films testifies to their lower susceptibility to oxidation compared to ZrN/SiN$_x$ films. As an example, the SEM micrographs of ZrN/SiN$_x$, CrN/SiN$_x$ and AlN/SiN$_x$ multilayered films with 5 nm/10 nm thickness ratio of elementary layers are represented in Figure 10. The lowest content of oxygen in

CrN/SiN$_x$ and AlN/SiN$_x$ films after air-annealing was revealed for such thickness ratio (Figure 8). Large corroded areas were observed on the surface of ZrN/SiN$_x$ coating (Figure 10a), while the surface of the CrN/SiN$_x$ coating is characterized by the presence of much smaller defects, which, however, do not cause any failure of the coating integrity (Figure 10b). The surface relief of the AlN/SiN$_x$ film is rather uniform (Figure 10c). Shallow, nanoscale blisters are only visible, with no sign of localized oxidation. These observations confirm the better oxidation resistance of AlN/SiN$_x$ multilayers, comparatively to CrN/SiN$_x$ and ZrN/SiN$_x$ systems.

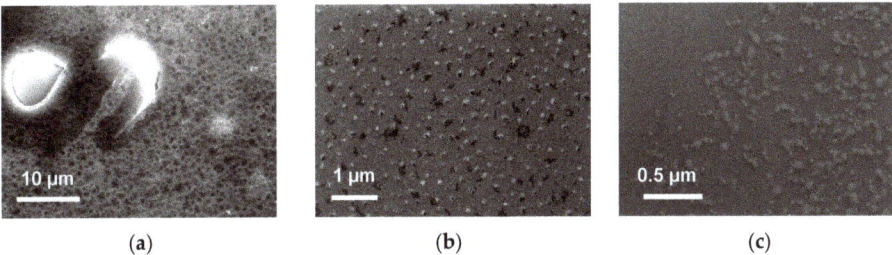

Figure 10. Top-view SEM micrographs of multilayered films with the ratio of thicknesses of MeN and SiN$_x$ elementary layers equal to 5 nm/10 nm after air annealing at 950 °C: ZrN/SiN$_x$ (**a**); CrN/SiN$_x$ (**b**); and AlN/SiN$_x$ (**c**). Note the different scale bar for the different images.

3.4. Discussion on the Comparative Oxidation Resistance of MeN/SiN$_x$ Multilayers and Me-Si-N Single-Layers

As pointed out in the Introduction, there exists two approaches for the synthesis of hard TMN coatings with enhanced phase stability during high-temperature air annealing, namely the formation of nanocomposite/amorphous Me-Si-N films or MeN/SiN$_x$ multilayered films. Therefore, it is rather interesting to compare their oxidation resistance. In general, the thermal stability of MeN$_x$ phase is determined by its decomposition into Me and N$_2$(g), which depends on the stoichiometry x = [N]/[Me]. In Zr-Si-N amorphous films, a worse thermal stability was reported for films with substoichiometric ZrN$_{x<1}$ phase [8]. Similar conclusions can also be made from the work of Abadias et al. on quaternary TiZrAl$_x$N$_y$ films [47]. However, the resistance to oxidation during air annealing of MeN$_x$ phase is also determined by its ability for crystallization of metal oxide (MeO$_x$) phases and the type of MeO$_x$ formed (dense solid vs. volatile oxides or passivating oxides) that impact coating morphology [4,9]. The present findings show that, despite CrN layers being substoichiometric (see Table 1), the oxidation resistance of CrN/SiN$_x$ multilayers (at least up to 950 °C) is higher than that of ZrN/SiN$_x$ multilayers (up to 860 °C).

Numerous works have been dedicated to study the oxidation behavior of Me-Si-N systems, including Zr-Si-N [7–10,48,49], W-Si-N [8–10], Ta-Si-N [8–10,50], Cr-Si-N [51], or Al-Si-N [8–10,52]. With increasing Si fraction, the microstructure typically evolves from bi-phase nanocomposites consisting of MeN$_x$ nanocrystals embedded in Si$_3$N$_4$ amorphous phase to X-ray amorphous phase. As an example, for Zr-Si-N coatings, the oxidation resistance rises with the increase of Si content, i.e. with increasing volume fraction of amorphous Si$_3$N$_4$ phase [6,48,49]. The best oxidation resistance is achieved for amorphous coatings, typically exceeding 1000 °C under air [9,52]. In the case of the Zr-Si-N system, the absence of mass gain was observed even up to the temperature of 1300 °C [9].

In a previous study [6], we compared the oxidation resistance of ZrSiN nanocomposite and ZrN/SiN$_x$ multilayered films. It should be pointed that the investigated films were relatively thin (thickness of ~300 nm) that resulted in a significant fraction of the oxidized layer. When using the same deposition conditions, the oxidation starts at 700–780 °C for ZrSiN nanocomposite films and at 860–950 °C for ZrN/SiN$_x$ multilayered films. This testifies to the advantage of the multilayered films. The lower degree of the compressive stress relaxation at the elevated temperatures of ZrSiN films, as compared to ZrN/SiN$_x$ multilayered films, was likely at the origin of film cracking, which deteriorates their oxidation resistance. The formation of cracks was indeed observed for nanocomposite and amorphous ZrSiN films despite reduction in oxygen penetration into the film with increasing Si content [6]. However, no cracks were observed for the ZrN/SiN$_x$ multilayers.

4. Summary and Conclusions

By means of magnetron sputter deposition technique, single-layer (monolithic) ZrN, CrN, AlN and Si$_3$N$_4$ films and MeN/SiN$_x$ (Me = Zr, Cr, Al) multilayered films with the different ratios of MeN layer thickness to SiN$_x$ layer thickness, namely 5 nm/2 nm, 5 nm/5 nm, 5 nm/10 nm and 2 nm/5 nm, were synthesized. The structure and phase composition of the films in as-deposited state as well as their stability under air-annealing in the 400–950 °C temperature range were studied.

TEM, XRR and XRD results disclose the formation of periodic MeN/SiN$_x$ multilayered structures characterized by uniform alternation of nanocrystalline MeN layers and amorphous SiN$_x$ layers with sharp and planar interfaces. In the case of ZrN/SiN$_x$ and CrN/SiN$_x$ systems, the MeN phase has (200) preferred orientation of the crystallites. In the case of AlN/SiN$_x$ film, the (002) preferred orientation was observed. For all multilayered films, the decrease in crystallites size of MeN phase occurs, and this phase becomes X-ray amorphous (as well as SiN$_x$ phase) when reducing the thickness of the corresponding elementary layer down to 2 nm.

The results of XRD, WDS and SEM analysis point that all reference monolithic films are subjected to significant oxidation level during high-temperature annealing. ZrN oxidizes into t-ZrO$_2$ at 550 °C, while CrN first decomposes into Cr$_2$N at 550 °C, followed by the formation of t-Cr$_2$O$_3$ phase at 700 °C. AlN and Si$_3$N$_4$ films appear more thermally stable, with lower oxygen uptake at 950 °C and no crystalline oxides detected. Compared to MeN single-layers, MeN/SiN$_x$ multilayers exhibit improved oxidation resistance due to the presence of amorphous Si$_3$N$_4$ layers: ZrN/SiN$_x$ multilayers start to oxidize at the temperatures of 780–860 °C, while CrN/SiN$_x$ and AlN/SiN$_x$ multilayered films are stable up to 950 °C. Further investigations at higher temperatures would be required to assess the upper temperature limit of their stability in air, and evaluate their performance comparatively to amorphous ternary Me-Si-N films, which are thermally stable and oxidation resistant up to ~1300 °C [9].

For ZrN/SiN$_x$ multilayered films, both the reduction of f_{MeN} fraction and increase in number of the bilayers in the film improve their oxidation resistance. However, ZrN/SiN$_x$ multilayered films are least thermally stable among the three studied systems. The minimum oxygen content (27.5 at. %) after air annealing is found for the ZrN/SiN$_x$ (2 nm/5 nm) film.

The CrN/SiN$_x$ and AlN/SiN$_x$ multilayered films are characterized by appreciably higher stability of their phase composition during air annealing. The oxygen content after annealing is in the 17.2–23.5 at.% (CrN/SiN$_x$ films) and 8.0–18.2 at.% (AlN/SiN$_x$ films) range. Both the passivating role of the chromium (or, especially, aluminum) oxides and the decrease in the compressive stresses during annealing (stress relaxation) are apparently the main reasons for the improvement of film properties. In contrast to ZrN/SiN$_x$ films, for CrN/SiN$_x$ and AlN/SiN$_x$ multilayered films, the decrease in MeN layer thickness down to 2 nm leads to certain deterioration of their oxidation resistance. This can be connected to the fact that the passivating role of the chromium or aluminum oxides in the surface layers of multilayered film is less effective at such a small thickness of elementary layer.

The CrN/SiN$_x$ and AlN/SiN$_x$ multilayered films with the thickness ratios of elementary layers of 5 nm/5 nm and 5 nm/10 nm are the most promising for practical applications at elevated temperatures (up to 950 °C). The absence of the explicit corrosion sites on the surface is evidenced for AlN/SiN$_x$ films. Among the studied samples, the lowest oxidation level was obtained for the AlN/SiN$_x$ (5 nm/10 nm) multilayer, for which the oxygen content after air annealing is only 8.0 at.%.

Author Contributions: Conceptualization, G.A. and V.U.; methodology, V.U. and A.J.v.V.; investigation, G.A., I.S., and J.H.O.; data curation, I.S. and S.Z.; visualization, I.S.; writing—original draft preparation, I.S.; writing—review and editing, G.A.; supervision, G.A.; and funding acquisition, V.U. All authors have read and agreed to the published version of the manuscript.

Funding: The work was partially supported by Belarusian Republican Foundation of Fundamental Research (Project F18MC-027).

Acknowledgments: The authors are grateful to Ph. Guérin for his technical assistance and scientific support on optimizing reactive growth process.

Conflicts of Interest: The authors declare no conflict of interest. The funders had no role in the design of the study; in the collection, analyses, or interpretation of data; in the writing of the manuscript, or in the decision to publish the results.

References

1. Milosev, I.; Strehblow, H.-H.; Navinsek, B. Comparison of TiN, ZrN and CrN hard nitride coatings: Electrochemical and thermal oxidation. *Thin Solid Films* **1997**, *303*, 246–254. [CrossRef]
2. Vaz, F.; Ferreira, J.; Ribeiro, E.; Rebouta, L.; Lanceros-Mendez, S.; Mendes, J.A.; Alves, E.; Goudeau, P.; Riviere, J.P.; Ribeiro, F.; et al. Influence of nitrogen content on the structural, mechanical and electrical properties of TiN thin films. *Surf. Coat. Technol.* **2005**, *191*, 317–323. [CrossRef]
3. Wen, F.; Meng, Y.D.; Ren, Z.X.; Shu, X.S. Microstructure, hardness and corrosion resistance of ZrN films prepared by inductively coupled plasma enhanced RF magnetron sputtering. *Plasma Sci. Technol.* **2008**, *10*, 170–175. [CrossRef]
4. Abadias, G.; Koutsokeras, L.E.; Siozios, A.; Patsalas, P. Stress, phase stability and oxidation resistance of ternary Ti–Me–N (Me = Zr, Ta) hard coatings. *Thin Solid Films* **2013**, *538*, 56–70. [CrossRef]
5. Barshilia, H.C.; Deepthi, B.; Arun Prabhu, A.S.; Rajam, K.S. Superhard nanocomposite coatings of TiN/Si$_3$N prepared by reactive direct current unbalanced magnetron sputtering. *Surf. Coat. Technol.* **2006**, *201*, 329–337. [CrossRef]
6. Saladukhin, I.A.; Abadias, G.; Uglov, V.V.; Zlotski, S.V.; Michel, A.; Vuuren, A.J. Thermal stability and oxidation resistance of ZrSiN nanocomposite and ZrN/SiN$_x$ multilayered coatings: A comparative study. *Surf. Coat. Technol.* **2017**, *332*, 428–439. [CrossRef]
7. Silva Neto, P.C.; Freitas, F.G.R.; Fernandez, D.A.R.; Carvalho, R.G.; Felix, L.C.; Tertoa, A.R.; Hubler, R.; Mendes, F.M.T.; Silva Junior, A.H.; Tentardini, E.K. Investigation of microstructure and properties of magnetron sputtered Zr-Si-N thin films with different Si content. *Surf. Coat. Technol.* **2018**, *353*, 355–363. [CrossRef]
8. Musil, J.; Vlček, J.; Zeman, P. Hard amorphous nanocomposite coatings with oxidation resistance above 1000 °C. *Adv. Appl. Ceram.* **2008**, *107*, 148–154. [CrossRef]
9. Musil, J. Hard nanocomposite coatings: Thermal stability, oxidation resistance and toughness. *Surf. Coat. Technol.* **2012**, *207*, 50–65. [CrossRef]
10. Musil, J. Advanced Hard Nanocoatings: Present State and Trends. In *Top 5 Contributions in Molecular Sciences*, 6th ed.; Avid Science; Telanga India: Berlin, Germany, 2020; pp. 2–65.
11. Abadias, G.; Michel, A.; Tromas, C.; Jaouen, C.; Dub, S.N. Stress, interfacial effects and mechanical properties of nanoscale multilayered coatings. *Surf. Coat. Technol.* **2007**, *202*, 844–853. [CrossRef]
12. Bobzin, K.; Brögelmann, T.; Kruppe, N.C.; Arghavani, M.; Mayer, J.; Weirich, T.E. Plastic deformation behavior of nanostructured CrN/AlN multilayer coatings deposited by hybrid dcMS/HPPMS. *Surf. Coat. Technol.* **2017**, *332*, 253–261. [CrossRef]
13. Chang, Y.-Y.; Weng, S.-Y.; Chen, C.-H.; Fu, F.-X. High temperature oxidation and cutting performance of AlCrN, TiVN and multilayered AlCrN/TiVN hard coatings. *Surf. Coat. Technol.* **2017**, *332*, 494–503. [CrossRef]
14. Contreras, E.; Galindez, Y.; Rodas, M.A.; Bejarano, G.; Gómez, M.A. CrVN/TiN nanoscale multilayer coatings deposited by DC unbalanced magnetron sputtering. *Surf. Coat. Technol.* **2017**, *332*, 214–222. [CrossRef]
15. Pogrebnjak, A.; Smyrnova, K.; Bondar, O. Nanocomposite Multilayer Binary Nitride Coatings Based on Transition and Refractory Metals: Structure and Properties. *Coatings* **2019**, *9*, 155. [CrossRef]
16. Lei, Z.; Liu, Y.; Ma, F.; Song, Z.; Li, Y. Oxidation resistance of TiAlN/ZrN multilayer coatings. *Vacuum* **2016**, *127*, 22–29. [CrossRef]
17. Xiao, B.; Li, H.; Mei, H.; Dai, W.; Zuo, F.; Wu, Z.; Wang, Q. A study of oxidation behavior of AlTiN-and AlCrN-based multilayer coatings. *Surf. Coat. Technol.* **2018**, *333*, 229–237. [CrossRef]
18. Kong, M.; Zhao, W.; Wei, L.; Li, G. Investigations on the microstructure and hardening mechanism of TiN/Si$_3$N$_4$ nanocomposite coatings. *J. Phys. D Appl. Phys.* **2007**, *40*, 2858. [CrossRef]

19. Wu, Z.; Zhong, X.; Liu, C.; Wang, Z.; Dai, W.; Wang, Q. Plastic Deformation Induced by Nanoindentation Test Applied on ZrN/Si$_3$N$_4$ Multilayer Coatings. *Coatings* **2018**, *8*, 11. [CrossRef]
20. Bai, X.; Zheng, W.; An, T.; Jiang, Q. Effects of deposition parameters on microstructure of CrN/Si$_3$N$_4$ nanolayered coatings and their thermal stability. *J. Phys. Condens. Matter* **2005**, *17*, 6405–6413. [CrossRef]
21. Soares, T.P.; Aguzzoli, C.; Soares, G.V.; Figueroa, C.A.; Baumvol, I.J.R. Physicochemical and mechanical properties of crystalline/amorphous CrN/Si$_3$N$_4$ multilayers. *Surf. Coat. Technol.* **2013**, *237*, 170–175. [CrossRef]
22. Hultman, L.; Bareño, J.; Flink, A.; Söderberg, H.; Larsson, K.; Petrova, V.; Odén, M.; Greene, J.E.; Petrov, I. Interface structure in superhard TiN-SiN nanolaminates and nanocomposites: Film growth experiments and ab initio calculations. *Phys. Rev. B* **2007**, *75*, 155437. [CrossRef]
23. Abadias, G.; Uglov, V.V.; Saladukhin, I.A.; Zlotski, S.V.; Tolmachova, G.; Dub, S.N.; Vuuren, A.J. Growth, structural and mechanical properties of magnetron-sputtered ZrN/SiN$_x$ nanolaminated coatings. *Surf. Coat. Technol.* **2016**, *308*, 158–167. [CrossRef]
24. Söderberg, H.; Odén, M.; Larsson, T.; Hultman, L.; Molina-Adareguia, J.M. Epitaxial stabilization of cubic-SiN$_x$ in TiN/SiN$_x$ multilayers. *Appl. Phys. Lett.* **2006**, *88*, 191902. [CrossRef]
25. Dong, Y.; Zhao, W.; Yue, J.; Li, G. Crystallization of Si$_3$N$_4$ layers and its influences on the microstructure and mechanical properties of ZrN/Si$_3$N$_4$ nanomultilayers. *Appl. Phys. Lett.* **2006**, *89*, 121916. [CrossRef]
26. Ghafoor, N.; Lind, H.; Tasnardi, F.; Abrikosov, I.A.; Odern, M. Anomalous epitaxial stability of (001) interfaces in ZrN/SiN$_x$ multilayers. *APL Mater.* **2014**, *2*, 046106. [CrossRef]
27. Parlinska-Wojtan, M.; Pélisson-Schecker, A.; Hug, H.J.; Rutkowski, B.; Patscheider, J. AlN/Si$_3$N$_4$ multilayers as an interface model system for Al$_{1-x}$Si$_x$N/Si$_3$N$_4$ nanocomposite thin films. *Surf. Coat. Technol.* **2015**, *261*, 418–425. [CrossRef]
28. Huang, L.; Chen, Z.Q.; Liu, W.B.; Huang, P.; Meng, X.K.; Xu, K.W.; Wang, F.; Lu, T.J. Enhanced irradiation resistance of amorphous alloys by introducing amorphous/amorphous interfaces. *Intermetallics* **2019**, *107*, 39–46. [CrossRef]
29. Mège-Revil, A.; Steyer, P.; Cardinal, S.; Thollet, G.; Esnouf, C.; Jacquot, P.; Stauder, B. Correlation between thermal fatigue and thermomechanical properties during the oxidation of multilayered TiSiN nanocomposite coatings synthesized by a hybrid physical/chemical vapour deposition process. *Thin Solid Films* **2010**, *518*, 5932–5937. [CrossRef]
30. Colin, J.J.; Diot, Y.; Guerin, P.; Lamongie, B.; Berneau, F.; Michel, A.; Jaouen, C.; Abadias, G. A load-lock compatible system for in situ electrical resistivity measurements during thin film growth. *Rev. Sci. Instrum.* **2016**, *87*, 023902. [CrossRef]
31. Abadias, G.; Koutsokeras, L.E.; Dub, S.N.; Tolmachova, G.N.; Debelle, A.; Sauvage, T.; Villechaise, P. Reactive magnetron cosputtering of hard and conductive ternary nitride thin films: Ti–Zr–N and Ti–Ta–N. *J. Vac. Sci. Technol. A* **2010**, *28*, 541–551. [CrossRef]
32. Simonot, L.; Babonneau, D.; Camelio, S.; Lantiat, D.; Guérin, P.; Lamongie, B.; Antad, V. In situ optical spectroscopy during deposition of Ag:Si$_3$N$_4$ nanocomposite films by magnetron sputtering. *Thin Solid Films* **2010**, *518*, 2637–2643. [CrossRef]
33. Abadias, G.; Chason, E.; Keckes, J.; Sebastiani, M.; Thompson, G.B.; Barthel, E.; Doll, G.L.; Murray, C.E.; Stoessel, C.H.; Martinu, L. Review Article: Stress in thin films and coatings: Current status, hallenges, and prospects. *J. Vac. Sci. Technol. A* **2018**, *36*, 20801. [CrossRef]
34. Parratt, L.G. Surface Studies of Solids by Total Reflection of X-Rays. *Phys. Rev.* **1954**, *95*, 359–369. [CrossRef]
35. Koutsokeras, L.E.; Abadias, G. Intrinsic stress in ZrN thin films: Evaluation of grain boundary contribution from in situ wafer curvature and ex situ X-ray diffraction techniques. *J. Appl. Phys.* **2012**, *111*, 093509. [CrossRef]
36. Abadias, G.; Ivashchenko, V.I.; Belliard, L.; Djemia, P. Structure, phase stability and elastic properties in the Ti$_{1-x}$Zr$_x$N thin-film system: Experimental and computational studies. *Acta Mater.* **2012**, *60*, 5601–5614. [CrossRef]
37. Patsalas, P.; Kalfagiannis, N.; Kassavetis, S.; Abadias, G.; Bellas, R.V.; Lekka, C.; Lidorikis, E. Conductive nitrides: Growth principles, optical and electronic properties, and their perspectives in photonics and plasmonics. *Mater. Sci. Eng. R* **2018**, *123*, 1–55. [CrossRef]
38. Musil, J. Flexible Hard Nanocomposite Coatings. *RSC Adv.* **2015**, *5*, 60482–60495. [CrossRef]

39. Xingrun, R.; Zhu, H.; Meixia, L.; Jiangao, Y.; Hao, C. Comparison of microstructure and tribological behaviors of CrAlN and CrN film deposited by DC magnetron sputtering. *Rare Met. Mater. Eng.* **2018**, *47*, 1100–1106. [CrossRef]
40. Ren, X.; Zhang, Q.; Huang, X.; Su, W.; Yang, J.; Chen, H. Microstructure and tribological properties of CrN films deposited by direct current magnetron sputtering. *Rare Met. Mater. Eng.* **2018**, *47*, 2283–2289. [CrossRef]
41. Khan, S.; Shahid, M.; Mahmood, A.; Shah, A.; Ahmed, I.; Mehmood, M.; Aziz, U.; Raza, Q.; Alam, M. Texture of the nano-crystalline AlN thin films and the growth conditions in DC magnetron sputtering. *Prog. Nat. Sci. Mater. Int.* **2015**, *25*, 282–290. [CrossRef]
42. Taurino, A.; Signore, M.A.; Catalano, M.; Kim, M.J. (101) and (002) oriented AlN thin films deposited by sputtering. *Mater. Lett.* **2017**, *200*, 18–20. [CrossRef]
43. Signore, M.A.; Taurino, A.; Valerini, D.; Rizzo, A.; Farella, I.; Catalano, M.; Quaranta, F.; Siciliano, P. Role of oxygen contaminant on the physical properties of sputtered AlN thin films. *J. Alloys Compd.* **2015**, *649*, 1267–1272. [CrossRef]
44. Riah, B.; Ayad, A.; Camus, J.; Rammal, M.; Boukari, F.; Chekour, L.; Djouadi, M.A.; Rouag, N. Textured hexagonal and cubic phases of AlN films deposited on Si (100) by DC magnetron sputtering and high power impulse magnetron sputtering. *Thin Solid Films* **2018**, *655*, 34–40. [CrossRef]
45. Mayrhofer, P.H.; Rovere, F.; Moser, M.; Strondl, C.; Tietema, R. Thermally induced transitions of CrN thin films. *Scr. Mater.* **2007**, *57*, 249–252. [CrossRef]
46. Lin, J.; Moore, J.J.; Wang, J.; Sproul, W.D. High temperature oxidation behavior of CrN/AlN superlattice films. *Thin Solid Films* **2011**, *519*, 2402–2408. [CrossRef]
47. Abadias, G.; Saladukhin, I.A.; Uglov, V.V.; Zlotski, S.V.; Eyidi, D. Thermal stability and oxidation behavior of quaternary TiZrAlN magnetron sputtered thin films: Influence of the pristine microstructure. *Surf. Coat. Technol.* **2013**, *237*, 187–195. [CrossRef]
48. Pilloud, D.; Pierson, J.F.; Marco de Lucas, M.C.; Alnot, M. Stabilisation of tetragonal zirconia in oxidized Zr–Si–N nanocomposite coatings. *Appl. Surf. Sci.* **2004**, *229*, 132–139. [CrossRef]
49. Chen, Y.-I.; Chang, S.-C.; Chang, L.-C. Oxidation resistance and mechanical properties of Zr–Si–N coatings with cyclic gradient concentration. *Surf. Coat. Technol.* **2017**, *320*, 168–173. [CrossRef]
50. Chen, Y.-I.; Gao, Y.-X.; Chang, L.-C. Oxidation behavior of Ta-Si-N coatings. *Surf. Coat. Technol.* **2017**, *332*, 72–79. [CrossRef]
51. Mikula, M.; Grančič, B.; Drienovský, M.; Satrapinskyy, L.; Roch, T.; Hájovská, Z.; Gregor, M.; Plecenik, T.; Čička, R.; Plecenik, A.; et al. Thermal stability and high-temperature oxidation behavior of Si–Cr–N coatings with high content of silicon. *Surf. Coat. Technol.* **2013**, *232*, 349–356. [CrossRef]
52. Musil, J.; Remnev, G.; Legostaev, V.; Uglov, V.; Lebedynskiy, A.; Lauk, A.; Procházka, J.; Haviar, S.; Smolyanskiy, E. Flexible hard Al-Si-N films for high temperature operation. *Surf. Coat. Technol.* **2016**, *307*, 1112–1118. [CrossRef]

© 2020 by the authors. Licensee MDPI, Basel, Switzerland. This article is an open access article distributed under the terms and conditions of the Creative Commons Attribution (CC BY) license (http://creativecommons.org/licenses/by/4.0/).

MDPI
St. Alban-Anlage 66
4052 Basel
Switzerland
Tel. +41 61 683 77 34
Fax +41 61 302 89 18
www.mdpi.com

Coatings Editorial Office
E-mail: coatings@mdpi.com
www.mdpi.com/journal/coatings

www.ingramcontent.com/pod-product-compliance
Lightning Source LLC
LaVergne TN
LVHW070608100526
838202LV00012B/594